Blockchain Technologies

Series Editors

Dhananjay Singh, Department of Electronics Engineering, Hankuk University of Foreign Studies, Yongin-si, Korea (Republic of)

Jong-Hoon Kim, Kent State University, Kent, OH, USA

Madhusudan Singh, Endicott College of International Studies, Woosong University, Daejeon, Korea (Republic of)

This book series aims to provide details of blockchain implementation in technology and interdisciplinary fields such as Medical Science, Applied Mathematics, Environmental Science, Business Management, and Computer Science. It covers an in-depth knowledge of blockchain technology for advance and emerging future technologies. It focuses on the Magnitude: scope, scale & frequency, Risk: security, reliability trust, and accuracy, Time: latency & timelines, utilization and implementation details of blockchain technologies. While Bitcoin and cryptocurrency might have been the first widely known uses of blockchain technology, but today, it has far many applications. In fact, blockchain is revolutionizing almost every industry. Blockchain has emerged as a disruptive technology, which has not only laid the foundation for all crypto-currencies, but also provides beneficial solutions in other fields of technologies. The features of blockchain technology include decentralized and distributed secure ledgers, recording transactions across a peer-to-peer network, creating the potential to remove unintended errors by providing transparency as well as accountability. This could affect not only the finance technology (crypto-currencies) sector, but also other fields such as:

Crypto-economics Blockchain
Enterprise Blockchain
Blockchain Travel Industry
Embedded Privacy Blockchain
Blockchain Industry 4.0
Blockchain Smart Cities,
Blockchain Future technologies,
Blockchain Fake news Detection,
Blockchain Technology and It's Future Applications
Implications of Blockchain technology
Blockchain Privacy
Blockchain Mining and Use cases
Blockchain Network Applications
Blockchain Smart Contract
Blockchain Architecture
Blockchain Business Models
Blockchain Consensus
Bitcoin and Crypto currencies, and related fields

The initiatives in which the technology is used to distribute and trace the communication start point, provide and manage privacy, and create trustworthy environment, are just a few examples of the utility of blockchain technology, which also highlight the risks, such as privacy protection. Opinion on the utility of blockchain technology has a mixed conception. Some are enthusiastic; others believe that it is merely hyped. Blockchain has also entered the sphere of humanitarian and development aids e.g. supply chain management, digital identity, smart contracts and many more. This book series provides clear concepts and applications of Blockchain technology and invites experts from research centers, academia, industry and government to contribute to it.

If you are interested in contributing to this series, please contact msingh@endicott.ac.kr OR loyola.dsilva@springer.com

More information about this series at http://www.springer.com/series/16276

James R. Reagan · Madhusudan Singh

Management 4.0

Cases and Methods for the 4th Industrial Revolution

 Springer

James R. Reagan
Endicott College of International Studies
Woosong University
Daejeon, Korea (Republic of)

Madhusudan Singh
Endicott College of International Studies
Woosong University
Daejeon, Korea (Republic of)

ISSN 2661-8338   ISSN 2661-8346   (electronic)
Blockchain Technologies
ISBN 978-981-15-6753-7   ISBN 978-981-15-6751-3   (eBook)
https://doi.org/10.1007/978-981-15-6751-3

This Springer imprint is published by the registered company Springer Nature Singapore Pte Ltd.
The registered company address is: 152 Beach Road, #21-01/04 Gateway East, Singapore 189721, Singapore

Introduction—Industry 4.0

What is the Fourth Industrial Revolution?

For the fourth time in recorded history, the way the world works is changing in ways that can only be called revolutionary. As with the first, second, and third industrial revolutions, these changes affect not only business but all humanity, transforming the way we work, play, and live.

But the *fourth industrial revolution*, which is the convergence of digital technologies to connect people and things in interdependent cyber-physical systems, differs from its predecessors in key ways.

This revolution, also known as *Industry 4.0*, is replacing old paradigms not only in business but in all aspects of daily life and is doing so with unprecedented speed. Rather than mere *innovations,* defined as improving an existing product or service in new and meaningful ways, the connected age is an era of *disruptions*—of not merely adding to existing markets, industries, and technologies, but destroying them, and replacing them with something completely new.

This book will explore 11 industries and how Industry 4.0 is affecting each:

- Agriculture,
- Automotive,
- Consumer,
- Energy,
- Financial services,
- Health care,
- Manufacturing,
- Media and entertainment,
- Retail,
- Transportation and travel.

A Social Impacts chapter will discuss the fourth industrial revolution's effects on society, and how the business sector can help ensure that those changes serve the greater good.

To finish, What's Next will explore the challenges and opportunities that lie ahead, including the fifth industrial revolution (Industry 5.0), in which humans and intelligent machines will work together; and how enterprises can prepare and position themselves for success in the new digital marketplace.

For the sake of brevity and clarity, we have paired each industry with only one of the technologies transforming it.

In the Agriculture chapter, for instance, we discuss how agribusiness is using *drones*, or remotely controlled flying robots, to inspect crops, plant seeds, determine soil composition, apply fertilizers and pesticides, and perform other tasks previously handled by humans. But farmers are also using the *data*, or computer-generated information, that drones collect to make decisions about how to extract the most value from their land and other resources; *autonomous vehicles*—vehicles that use artificial intelligence to operate themselves—such as self-driving tractors to plow fields and bale hay; and other technologies. Digital technologies promise to free agriculturalists from routine tasks so they may focus on their business strategies and improve the precision of their farming practices—watering only where fields are dry, for instance, or harvesting only the fruit that is ripe.

As you read, remember that all the technologies we will discuss throughout this course may affect any of the industries covered in these pages. Retailers use drones, too, to deliver products; nearly every business uses data; the transportation industry also uses autonomous vehicles.

In addition to the analysis of an industry and one technology affecting it, each chapter also will contain:

- A *case study*, for example, of how T4IR is changing one specific enterprise within the spotlighted industry,
- A short list of terms introduced in the chapter, for you to learn, and
- Questions about what you've read.

Before we proceed, however, let's take a step back to examine the fourth industrial revolution in the context of time and events, and where it fits into the overall picture.

The Four Industrial Revolutions

For most of human history, people worked solely with their hands. Every item people used—clothing, tools, cooking implements, musical instruments, weapons —was made individually by someone, using raw materials collected or cultivated from the Earth. Our ancestors hunted and gathered all their food until about 10,000 years ago when people began farming, and domesticated animals helped with jobs requiring additional strength, endurance, and speed.

1. **The first industrial revolution** (mid-18th–mid-19th centuries) used the steam engine to mechanize manufacturing processes—something that had never been done before. As a result, factories sprang up, producing goods more quickly and cheaply than could be done by hand, enabling more people to procure and use the goods. These factories, although fairly small, also employed more paid workers, which raised the overall standard of living.
2. **The second industrial revolution** (early 20th century) used electricity to speed manufacturing processes even more, aided by the *assembly line*, pioneered by US automaker Henry Ford, in which items move progressively from worker to worker, each of whom completes a single step in product assembly. This method of manufacturing greatly increased productivity and allowed automobiles and other complex items to be mass-produced for wide distribution.
3. **The third industrial revolution** (late-20th century) occurred with the invention of the Internet, which enabled goods and services to be produced, marketed, and consumed globally and introduced digital technology worldwide.
4. **The fourth industrial revolution** (early 21st century), now underway, connects people and things with digital technologies including *artificial intelligence*, which uses computers to perform tasks that previously required human intelligence such as visual perception, speech recognition, decision making, and language translation; the *Internet of Things*, in which inanimate items communicate with humans and one another, and *cloud computing*, the delivery of computing services over the Internet to enable all this communication; and many others.

The Fourth Industrial Revolution

Cars, trucks, buses, and trains that drive themselves. Homes and offices that control themselves, using thermostats that adjust automatically; refrigerators that order groceries; doors that lock and unlock themselves as we leave and approach.

Factories whose robots assemble and manufacture goods alongside or instead of humans, make repairs, place and fill orders, and carry out many other functions autonomously. Medical technologies let us track and monitor our health, diagnose illnesses, and even treat ourselves. We can spend, save, and invest digital money without opening a bank account. Retail stores let us buy on the spur of any moment, from wherever we happen to be.

The "connected age" is bringing all this and more to people and organizations around the world at lightning speed. While the first industrial revolution took about one hundred years to manifest, Industry 4.0 is happening so quickly that most businesses are challenged just to keep pace. Even early adopters who have invested amply in technologies may find that, in a few years, they have fallen behind as new inventions and improvements render what they have obsolete.

Consumers, however, are not behind the curve. Instead, they are pushing ahead. Empowered in new ways by their mobile phones, tablets, and other computing devices, they are making new demands for around-the-clock, instant access to goods and services, customization and personalization of those goods and services, and novel, stimulating experiences.

The New Business Paradigm

The providers of these goods and services—primarily businesses—will operate differently than in the past. Tomorrow's businesses will be fully *customer-centric*, prepared to accommodate customers' demands, to innovate in the instant as those demands change, and to anticipate their future needs. Digital technologies will enable this new approach.

Being fully digital is already becoming imperative for enterprises in all industries. Business leaders will need to shift not only their mindset but also their business models. Every business will need a digital strategy, which will include deciding:

- Which technologies to use,
- The best ways to use technologies,
- How to secure digital data, networks, and systems,
- How to assist customers with the use of their technologies.

To get there, businesses will need to adopt new *paradigms*, or models. In Industry 4.0, they will move away from the linear *value chain*, which is the process by which a company turns raw materials into saleable goods, including production, marketing, and after-sales service. Rather than a top-down approach, the new model entails coordination, collaboration, and partnerships in *business ecosystems*, or networks of interconnected entities (including the customer) all working together to create or enhance a product or service.

When you have completed this course, you should understand the concepts and technologies driving the fourth industrial revolution; how the interconnection of digital technology, people, and things affects industries and societies; the challenges and opportunities this shift presents to the public and private sectors, and how businesses can adapt and change to increase their chances of success in the "connected age."

Germany's Industrie 4.0 program

The government of Germany wants to lead the way into the fourth industrial revolution. With the goal of establishing the nation as a world leader in manufacturing, government ministries have established Industrie 4.0, a strategic initiative funding technological research, forming industry networks, and standardizing technologies.

The Technologies

Funded by the Ministry of Education and Research and the Ministry for Economic Affairs and Energy, Industrie 4.0 envisions improvements in manufacturing processes using such technologies as:

- **Data**: Collecting and analyzing data can help improve quality control by identifying weaknesses and flaws in the manufacturing system and finding solutions.
- **Autonomous vehicles**: Logistics vehicles operate themselves, moving goods automatically and intelligently from one production point to the next.

- **Cyber-physical systems**: In tomorrow's fully automated "smart" factories, machines oversee production for increased productivity, less downtime, and lower personnel costs.
- **Robotics**: Equipped with sensors, cameras, and digital connections, automated, autonomous robots can work on the factory floor, performing many tasks previously accomplished by humans without human error or fatigue.
- **Production line simulation**: *Digital twins*, or virtual prototypes, allow engineers to safely test production line designs for efficiency and effectiveness before they are implemented.

The Benefits

The government isn't the only investor in Industrie 4.0. A number of research centers, a consortium of industry stakeholders, and private industries are collaborating on the initiative.

The potential for gain is high: One study estimated that Industrie 4.0 could generate 79 billion euros' growth in the country by 2025 in six sectors: chemical engineering, automotive, mechanical engineering, IT and communication, electrical engineering, and agriculture.

- **More effective and efficient workplaces**. Factories can operate more smoothly and at lower cost using connected technologies, as can businesses in many research and developments in one sector benefits all.
- **Enhanced reputation as a world industrial center**. Manufacturing already makes up a large portion of the German economy, making it one of the most competitive nations in the world for industry. By looking to the future, the country hopes to maintain that status.
- **A competitive technological edge**. Industrie 4.0 is a private–public partnership, with research and development funded by the government as well as business dollars. Technologies developed and enhanced under the initiative may ultimately benefit these businesses as well as the German economy.
- **A boosted national economy**. Increased productivity, improved customer services, a workforce trained to perform in the digital age, and standard-setting technology products may stimulate investments in German business, industry, and infrastructure, creating jobs and revenues and enhancing the quality of life. The added gross value nationwide generated by Industrie 4.0 is expected to average 1.7 percent per year.

Case Study—Germany Industrie 4.0

(see Appendix 117–119)

Challenges and Lessons Learned

Security. Securing data is difficult enough, and the stakes are high for privacy and proprietary information. Securing all the devices used in a smart factory presents a unique set of challenges, with failure potentially causing breakdowns and even danger to human safety.

Infrastructure. Particularly in rural areas, Germany suffers from slow and even spotty Internet service. Without fast broadband, engineers and manufacturers may not even realize the possibilities the digital age offers to manufacturing.

Standardization. Although this is a goal of Industrie 4.0, many technology products are still available using many different platforms and interfaces. Smaller companies and suppliers, in particular, will have a harder time converting to digital if they have to continually buy new technologies and train workers in how to use each new software product.

Glossary of Terms

Fourth industrial revolution: The convergence of digital technologies to connect people and things in interdependent cyber-physical systems

Industry 4.0: Another name for the fourth industrial revolution

Innovation: Inventing a new product or service or improving an existing product or service in new and meaningful ways

Disruption: Not merely adding to existing markets, industries, and technologies, but destroying them, and replacing them with something completely new

Drone: A remotely controlled flying robot

Data: Computer-generated information

Autonomous vehicles: Vehicles that use artificial intelligence to operate themselves

Assembly line: A manufacturing method pioneered by US automaker Henry Ford in which manufactured items move progressively from worker to worker, each of whom completes a single step in product assembly

Artificial intelligence: A form of digital technology that uses computers to perform tasks that previously required human intelligence such as visual perception, speech recognition, decision making, and language translation

Internet of Things: Digital technologies that connect humans to inanimate things and things to one another

Cloud computing: The delivery of computing services over the Internet

Paradigm: A model or pattern

Value chain: The process by which a company turns raw materials into saleable goods including production, marketing, and after-sales service

Business ecosystem: A network of interconnected entities (including the customer) all working together to create or enhance a product or service

Digital twin: A virtual prototype that allows engineers to safely test factory production line designs for efficiency and effectiveness before they are implemented

Questions

1. Describe each of the three previous industrial revolutions. How does the fourth differ from these? How is it similar?
2. Name four technologies essential to business in the "connected age."
3. "The customer is always right." Is this still true in the fourth industrial revolution? What role does the customer now play, and how can businesses adapt?
4. What questions do business leaders need to consider as they develop their digital strategies?
5. What are some benefits of Germany's "Industrie 4.0" program? What are the challenges?

Contents

Chapter 1
Agriculture Revolution

1.1 Agriculture and Drones

If data is the "new gold," then drones are the gold miners and refiners. Originally developed for military surveillance and operations, drones, which are flying robots that are remotely controlled or fly autonomously using flight plans controlled by software, now serve as data collectors in a variety of industries, providing eyes, ears, and even hands in places where humans cannot easily go, and doing it more rapidly, easily, and cheaply.

Drones come in sizes as large as an airplane or as tiny as an insect. Equipped with sensors, cameras, and compartments or arms for carrying payloads, or items to deliver, drones are used for an increasing number of tasks and in a growing number of industries and sectors, including:

- Search and rescue,
- Law enforcement,
- Infrastructure inspection and repair,
- Building inspections,
- 3D mapping,
- Solar and wind turbine monitoring,
- Cell tower inspection,
- Photography,
- Construction safety monitoring,
- Cargo delivery.

There are even aquatic drones that operate underwater to find and help clean up spills, garbage, and shipwrecks.

J. R. Reagan and M. Singh, *Management 4.0*, Blockchain Technologies, https://doi.org/10.1007/978-981-15-6751-3_1

1.1.1　Drones and Agriculture

One industry adopting drones on a widespread basis is agriculture. Small farmers and big agri-business alike are using drones to:

Scan fields for soil analysis. Drones scan crop fields, flying low, using sensors, or devices that detect and respond to physical stimuli, and cameras to create three-dimensional maps with information about soil composition, nutrients (and deficiencies), moisture, and more.

Plant crops. The aircraft drop pods containing seeds and fertilizer into the soil in precise locations and at pre-set intervals.

Apply pesticides and fertilizers. Using ultrasonic echoing, a method of measuring distance by emitting sound waves and measuring the time it takes for the sound to echo back, and lasers, drones detect altitude shifts and topography as they spray or dust crops, continually adjusting for even, efficient, more environmentally friendly coverage.

Monitor crops and livestock. Images and sensor information enable farmers to view crop health over a vast area, determine nutrient and water needs, know when crops are ready for harvest, and locate and monitor livestock in real time.

Create GPS maps. The data that drones gather, once compiled and analyzed, can help farmers create detailed maps of their croplands showing soil characteristics, yields, and other qualities inch by inch. This detailed data helps with precision farming, also known as site-specific crop management, a method of agriculture that uses information technology to determine what individual crops need and guide their decisions for the next planting season.

1.1.2 The Four Industrial Revolutions: Agriculture

When humans first began cultivating crops about 10,000 years ago, they did everything by hand or with the help of animals. Only in the last several hundred years have people used machines to make the task easier and faster.

1. **Mechanization (eighteenth century)**: Two inventions, the seed drill, which cut furrows in fields and deposited seeds, and the cotton gin, a hand-operated device (later powered by steam) that separated cotton fiber from seed 50 times faster than separation by hand, demonstrated how mechanical devices could make farming much easier and more efficient.
2. **Industrialization (mid-19th-early twentieth centuries)**: As railroad and steamship lines expanded, new markets emerged for agricultural products. Gasoline-powered tractors replaced steam-powered ones as well as, still in use, plows pulled by draft animals, and the combine harvester, which both cut and threshed grain crops, reduced the labor needed to harvest one hectare of wheat (100 acres) from 37 h to just 6.25 h.
3. **Technology (late twentieth century)**: US researchers spliced a gene from one organism into another, sparking the age of genetic engineering, the manipulation of genes to make plants and animals stronger, more disease-resistant, and more productive. Also, farmers began using computers to track, analyze, and plan their crop and livestock management.
4. **Digitization**: Drones gathering data enable farmers to pinpoint and treat problem areas and improve crop yields, helping to feed the world's growing population from a shrinking land base.

1.1.3 How Agriculture Works Today

Precision agriculture has been practiced for decades. In the 1980s, farmers began using compasses to map their lands and make fertility notes, and variable-rate applicators to spread fertilizers based on those maps.

The advent of global positioning systems (GPS) for navigation helped make those maps more precise. However, farmers still must walk their lands to inspect their crops. Spot-checking, or inspecting samples gathered at random, is the most common method used to get information about crop health, but it is time-consuming and not accurate across an entire field.

Some farms use piloted aircraft and satellites to capture data with near-infrared light, a form of radiation invisible to the human eye, and red, green, and blue (RGB) cameras to assess their fields, but these expensive solutions are impractical for all but the largest operations.

Livestock ranchers must ride their properties using motor vehicles and horses to locate their herds and drive them to different pastures or to shelter or processing facilities. Problems with health or birthing may be missed until it is too late.

To apply fertilizers, herbicides, and insecticides, farmers must use heavy machinery to lift their containers, which can weigh 500 lb, and apply the treatments using trucks or crop dusters, piloted aircraft equipped to spray crops from the air. Crop dusting is a dangerous profession, and pesticide drift, the settling of pesticide droplets or dust off-site, can cause health problems to those exposed.

And at harvest time, crops get harvested all at once, when the majority of the plants are ripe, meaning that some are gleaned before they are ready and others when they are past their prime. What is more, problems with disease, infestations, or watering may not be discovered until it is too late to save the stressed plants. On the farm, timing is everything, and "too little, too late" can be expensive and even devastating.

The Digital Revolution

For farmers using them, drones are already changing the way agriculture gets done.

Instead of making time-consuming, inaccurate spot checks, they can send a drone to survey their fields quickly and thoroughly, using the data to create precise maps showing their fields in detail: which diseases or pests are where, which areas need more or less water, which plants lack which nutrients, and more.

Instead of having to spray or dust their fields using trucks or piloted aircraft, they can send a drone to problem areas without risk to a human pilot's health or safety. The drones can fly close to the ground and apply chemicals so precisely that the risk of drift is also greatly reduced.

Rather than having to drive or ride in search of their grazing livestock, they can send a drone out to search for the animals and see on a digital map exactly where they are, whether any are missing (and find those animals, too), and whether any are injured or ill. Some are using drones to herd their livestock to their next destination, as well.

Drones are also proving valuable for maintaining the farm's physical assets. Rather than having to walk, ride, or drive the perimeter of the farm or ranch to check for fencing or other structures in need of repair, the farmer can send a drone to photograph them.

1.1.3.1 What's Next

As drones become cheaper, and the sensors, cameras, and software they use become more sophisticated, and they will almost certainly assume more and more of the tasks that farmers and their employees conduct today.

In tandem with autonomous vehicles, including tractors, drones have demonstrated the ability to plant, grow, and harvest fields of crops, even working in fleets, without human intervention—sweeping through vineyards and picking only the grapes that are perfectly ripe, for instance, leaving the rest for another time.

In the future, using artificial intelligence, drones may detect when pest control or fungicides are needed and schedule themselves, based on weather forecasts, to apply them at the best time, day or night.

And they may someday be able to act as virtual sheepdogs, tracking and guiding livestock without human direction, using pre-programmed schedules and maps.

1.1.3.2 Obstacles and Challenges

Revolutions bring challenges. Laws and regulations must change; problems arise; mindsets must shift as individuals and societies adjust to the new paradigms dramatic change requires. Some of those obstacles and challenges regarding the use of drones and other technologies in agriculture include:

Regulations. Not every country has laws or regulations restricting the use of drones, but an increasing number do. Laws tend to center on privacy and safety concerns, setting rules about how low drones may fly over someone else's privately owned property, for instance, or requiring their on-the-ground pilots to be able to see them at all times.

Having to keep these aircraft within sight may not be much of a hindrance, while the technology and software are still too immature to allow their operating independently. As artificial intelligence and other technologies improve, however, drones will be able to perform many tasks on their own, even while the farmer sleeps at night. At that point, laws may need to change to reflect these new abilities to avoid collisions with birds or planes, hitting structures such as cell towers, or spraying chemicals on people or on the wrong fields.

Security. As with all connected devices, cybersecurity is a concern. Agriculturalists must take responsibility to ensure that any digital technologies they use are safeguarded against breach or control by any unauthorized entity.

Mindset. Someday, it is thought, UAVs in communication with other connected devices such as autonomous tractors will be able to do much of the labor of farming more cheaply, efficiently, and effectively than farmers can do. This change could minimize the tedium, uncertainty, labor, and expense involved in the industry today and boost productivity to feed growing populations on the Earth's dwindling arable lands.

Rather than spending their time driving tractors, baling hay, feeding livestock, and planting and harvesting crops—or hiring workers to perform these tasks—farmers will use fleets of drones and other automated equipment, freeing them to analyze and act as a conductor to the orchestra of devices working the fields. Tomorrow's farmers will be data analysts, programmers, and drone pilots. To get there may require a shift in mindset, however. Farmers are often traditional in their views, with practices handed down through generations. They may be slow to embrace change. Also, the investments they have already made in costly equipment may make them hesitant to shift gears. On the other hand, the transformation of agriculture into a digitally oriented profession may attract younger people to the job.

Skillsets. As their neighbors and competitors educate themselves about drones and other new technologies and adapt to their new, tech-centered roles, farmers who cling to the old ways may risk being left behind—their yields less than optimal, their labor costs unnecessarily high. To survive will require foresight and planning, and the acquiring of new skills—starting now.

1.2   Case Study—Hands-Free Hectare

(see Appendix 147–149)

Researchers in England used drone technology embedded in agriculture machinery to drill, sow, tend, and harvest without ever stepping foot on the soil. Hands-Free Hectare in 2017 produced the world's first crops grown completely with autonomous technologies, using drones, sensors, computers, and a modified tractor and combine.

In the near future, the researchers say, all farms will operate this way.

1.2.1   The Technologies

The project, at Harper Adams University in Shropshire, England, used lightweight equipment so as not to compact the soil, which can cause water to run off instead of soaking in and hinder root growth.

With £200,000 from the British government and a private investor, the researchers equipped a small tractor and a 25-year-old combine harvester with a sprayer boom, a seed drill, and drone technology—cameras, sensors, lasers, GPS, and more. Rather than buying expensive autonomous machines, the teams opted to modify the existing equipment, in part because of their small budget, but also to show that farmers can use the techniques without breaking their bank accounts.

1.2.2 The Benefits

The drone software not only turned the steering wheel, turned the spray nozzles on and off, and raised and lowered the drill, but also told the machines which route to take and where and when to turn. In its first year, the tractor first applied herbicide and then, over six hours, paused at programmed points to drill, plant seeds, and add fertilizer.

Drones equipped with infrared sensors and a robot "scout" monitored soil moisture and crop conditions, sending images to the researchers to view on their computers. Drones also collected crop samples for the researchers, who inspected them to determine whether they were ready for harvest.

Greater efficiency. The stated goal of the project was to prove that there is no technological reason why fields cannot be farmed without humans working the land. The Hands-Free Hectare yielded a good crop without anyone's even setting foot on the field—theoretically freeing up farmers to focus on the business, ultimately improving productivity and profits.

Higher crop yields. Using technology enhances precision farming techniques, resolving issues such as a lack of important nutrients or an infestation of pests before they become problems.

Less waste. Being able to harvest crops when they are ripe rather than on the farmer's schedule could reduce spoilage and waste.

Cost savings. Using machines to perform manual tasks can reduce the need for human labor as well as their costs.

Convenience. Being able to send out machines saves time and—may get the job done—no matter what the weather.

Flexibility. Not having to operate equipment manually or be physically on the scene means that, instead of being limited to using one tractor or harvester at a time, farmers could have several working for them, all in the same field or in different areas of the farm, all at once or at different times.

1.2.3   Challenges and Lessons Learned

The technology. Trying to keep costs as low as possible, the researchers retrofitted the existing digital technology onto analog equipment. This meant programming software to, for instance, guide tractors so that they would travel in a straight line rather than veering around objects such as rocks, which could cause damage to fragile plants. Adapting technologies saved costs over buying expensive autonomous tractors and other equipment, but it also resulted in a lot of trial-and-error and continual fine-tuning to improve accuracy and performance.

Lower crop yields. Yields after the first year were lower than a hectare likely would have produced if farmed using conventional methods, but harvests increased in subsequent seasons as the technologies improved—with, for example, seeds being planted more accurately.

Time and money. The amount of time and money spent farming the Hands-Free Hectare was much higher in the first year than if researchers had used conventional methods. Some tasks were more difficult, as well. For instance, monitoring fields remotely was said to be more difficult than checking them in person might have been. This was expected to improve with the technologies.

1.3   Glossary of Terms

Drones   Flying robots that are remotely controlled or fly autonomously using flight plans controlled by software.

Payloads   Items that drones deliver.

Sensors   Devices that detect and respond to physical stimuli.

Ultrasonic echoing A method of measuring distance by emitting sound waves and measuring the time it takes for the sound to echo back.

Precision farming   A method of agriculture that uses information technology to determine what individual crops need and guide their decisions for the next planting season, also known as site-specific crop management.

Spot-checking   Inspecting plant samples gathered at random to get information about crop health.

Near-infrared light A form of radiation invisible to the human eye, often used for measurement.

Crop dusters   Piloted aircraft equipped to spray crops from the air.

Pesticide drift   The settling of pesticide droplets or dust off-site.

1.4   Questions

1. Name five uses for drones, imagining how they might affect a relevant industry or sector.
2. How are drones being used in agriculture? Name three, explaining how they can benefit farming. Can you think of other ways farmers might use them that are not included in this chapter?
3. How has each of the three previous industrial revolutions changed agriculture?
4. What are some of the changes Industry 4.0 promises to bring to farming and farming practices? Detail three practices or techniques used in farming that may change with the use of digital technologies.
5. Using the Hands-Free Hectare as an example, discuss how connected digital technologies might benefit agriculture.
6. What challenges or obstacles face farmers who might want to "go digital"? What changes might they need to make to their business model to keep up? What challenges face the industry as a whole?

Chapter 2
Automotive Evolution

2.1 Automotive and Artificial Intelligence

In the very near future, cars will drive themselves—and us. *Autonomous vehicle (AV)* technologies are not just putting robots in the driver's seat but are doing away with that seat altogether. In this new world, the motor vehicles *are* the robots, navigating highways and streetscapes efficiently and safely with little or no human assistance and, after dropping off their passengers, even parking themselves.

Such a large-scale transportation transformation will affect daily life and work throughout society and will require a profound shift in every part of the automotive industry, including:

- Design,
- Manufacturing,
- Sales,
- Service,
- Insurance.

That change entails a digital mindset, driven by a digital strategy.

J. R. Reagan and M. Singh, *Management 4.0*, Blockchain Technologies,
https://doi.org/10.1007/978-981-15-6751-3_2

2.2   The Four Industrial Revolutions: Automotive

From the invention of the wheel in prehistoric times, humans have used technology to transport themselves and their material possessions farther and faster than they could go on foot.

1. **Mechanization (eighteenth century)**: French military engineer Nicolas-Joseph Cugnot designed and built the first automobile in the late 1770s using an engine he also designed—one propelled by steam. His invention did not succeed, but it prefigured the steam engine and, eventually, the invention of the automobile, probably by the German Karl Benz around 1885.
2. **Industrialization (early twentieth century):** American businessman Henry Ford invented the *assembly line,* a method of production that moves items along a line of workers who perform a progressive series of repetitive tasks. This revolutionary method cut the time required to make an automobile from 12 to 2 h and 30 min.
3. **Technology (mid-to-late twentieth century)**: Electrification not only of factories but of the vehicles themselves and globalization of production accelerated advances in automobile technologies, including *advanced driver-assistance systems (ADAS)*, systems to aid drivers in safely operating vehicles. Using cameras, radar, and sensors, cars today can see around driver blind spots, warn of impending collisions and brake automatically, use cruise control becoming to set and maintain traveling speed, and more.
4. **The connected age**: Connected digital technologies enabling vehicles to communicate with one another, to gather information from other computing devices, and to analyze and act on data will result in vehicles' autonomously operating without the need for human assistance.

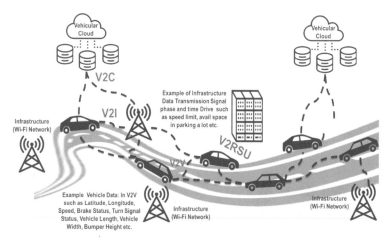

The connected age vehicles

2.3 The Technologies

Fully autonomous cars use computers to orchestrate and respond to information coming from sensors placed all over the car. These sensors include:

- Cameras,
- Motion and range detectors that use *ultrasonic* (sound waves with frequencies undetectable by humans), radar (high-frequency electromagnetic waves), and *lidar* (pulses of light) technologies,
- And, for positioning and mapping, *gyroscopes,* which measure orientation, and *accelerometers*, which measure motion.

The data these sensors collect not only helps guide each car efficiently and safely to its destination, but also improves the software's accuracy for other vehicles using it, updating such information as speed limits, road construction, and traffic conditions.

AVs also use connected digital technologies including the *Internet of Things* (*IoT*), digital technologies that connect humans to inanimate things and things to one another; *cloud computing,* which is the delivery of computing services over the Internet; and, most crucial, *artificial intelligence*, which enables vehicles to process data and make the driving decisions that humans would otherwise make.

2.4 The Benefits of AVs

More than 1 billion cars and trucks occupy the planet's roadways today, and that number is expected to double by 2040, the World Economic Forum reports. Congestion, fatalities, and climate change are some of the detrimental effects of all those vehicles. Motor vehicles, including cars, trucks, ships, and airplanes, produce 27% of greenhouse gas emissions, the second-highest contributor next to electricity production, according to the US Environmental Protection Agency. Traffic injuries were among the ten leading causes of death in 2015, killing 1.3 million people that year, the World Health Organization reported.

Autonomous vehicles promise to reduce the harm motor vehicles cause, benefiting individuals, businesses, and societies in myriad ways:

- Because most accidents are caused by human error, AVs are expected to reduce them greatly, and perhaps eliminate traffic deaths altogether—a goal known as *"Vision Zero."*
- Because AVs communicate with one another, stoplights and traffic jams could become a thing of the past, making trips faster and more efficient. Add the fact that cars park themselves, and travel time could decrease by as much as 40%.
- Because travelers would be free to perform other tasks while on the road, SDVs could also improve productivity, recovering up to one billion hours per day lost during commutes worldwide—about twice the time it took to build the Great Pyramid of Giza.
- Because AVs will operate more efficiently and also facilitate *ride-sharing*, or the sharing of a vehicle as it travels from one place to another; *platooning*, or the grouping of vehicles such as trucks together in a coordinated traffic stream; and *on-demand services,* or the use of a vehicle on demand for specific tasks, they are expected to lower fuel consumption by up to 40%.
- Because businesses won't need to pay drivers, they could slash their costs as much as 60%, depending on the business. Using less fuel would enable further savings: trucking companies spend about $90 billion per year on diesel fuel, their second-biggest expense next to driver pay, or about 20% of operating costs.

The lion's share of these benefits, especially at first, will go to urban and suburban areas, particularly in developed nations where rules of the road are clearly demarcated and well-enforced, and where the infrastructure is more sophisticated. Of cities surveyed by the Boston Consulting Group, 90% said they expected to have at least one shared-AV fleet operating within their borders by 2025.

Cities are natural environments for AV use, in part because increasing densities make driving a hassle and parking expensive. People worldwide continue to migrate to urban settings: although just three in ten people lived in cities 50 years ago, today half the world's people do, and that number is quickly growing. Here, drivers struggle with congested streets and the hassle and expense of parking, vehicle exhaust pollutes the air, and parking lots take up valuable real estate. While public transportation serves many, it is not always convenient, especially for helping people travel that *last mile* between home and a train, bus, or subway stop.

Driverless vehicles can resolve most, if not all, these problems, transporting riders—on demand—to public transportation or even into the city. Once there, these vehicles autonomously self-park (after locating an available spot in advance), and they also fit into more compact spaces since no room is needed for a human to exit the car.

Even more dramatic, AVs may eliminate the need for city dwellers and suburb dwellers to own a car at all, saving them money on payments and maintenance as well as insurance. Many cities may ban privately-owned vehicles altogether in an effort to reduce congestion, noise, pollution, and the need for such infrastructure as parking lots and garages. In another BCG survey, 84% of city policymakers said they expect such a ban in their cities by 2030. Oslo, Norway, planned to ban private cars from its city center by 2020, a prohibition expected to cut greenhouse gas emissions in half.

2.5   Challenges to AV Adoption

The very features that make cities ideal for AV use also pose barriers to their implementation. While traveling at slower speeds in city traffic makes fatal or serious accidents less likely, AV developers face special challenges programming vehicles to deal with the crowding, frequent route changes due to construction and other activities, and unpredictability that urban streets present. Cities must also find ways to integrate ridesharing and *robo-taxis,* or driverless taxicabs, into their public transportation systems for optimal public mobility at affordable prices.

Concerns abound about the technology, from not only potential users but also technology developers, policymakers, those in peripheral sectors such as trucking and insurance, and automotive industry *incumbents*, or those in business today. Obstacles to a fully self-driving world include:

- **Cost**. The first autonomous features, including *highway autopilot with lane-changing* and *urban autopilot*, both of which allow vehicles to drive themselves with human attention or assistance, were expected to add $5000 to $6000 to the price of a new car. For this reason, these features would likely be offered on higher-priced premium vehicles. To gain traction with the general public, the cost of the technology would need to decrease.
- **Safety**. Although AVs are predicted would reduce traffic accidents and even, perhaps, eliminate deaths from vehicle-related injuries, consumers have not been certain. Half of those surveyed in a BCG poll said that they would feel unsafe in a self-driving vehicle, and most parents said they would not let their child ride alone in one. Automakers will need to ensure that, in the competitive rush to market, they are not selling AVs before they are demonstrably safe. This entails working with the public sector as well as technologists on exhaustive pilot testing.
- **Security.** News reports of *white-hat*—"good"—hackers' gaining control of conventional vehicles have sounded alarm bells: What if bad actors remotely commandeered AVs and used the vehicles to cause harm? Current computerized vehicles operate using a *CAN bus*, an in-vehicle network that controls onboard *electronic control units* (*ECUs*). Hackers gaining control of one of these units could access them all and control an entire vehicle. The danger multiplies when sensors, controllers, and *telematics*, or "black boxes" recording vehicle-use data, transmit data to one another, the AV, automakers, insurers, and software manufacturers.
- Attacks are not the only cybersecurity concern. Many consumers worry about the security and privacy of their personal data, including their whereabouts, Internet use, personal identifying information, and other information. Although no system is completely secure from intrusions, automakers must work with others in the industry as well as throughout the private and public sector to spot threats, develop strategies to defend their vehicles and systems, respond in a coordinated fashion, and recovery quickly and completely should a breach occur. Along these lines, the automotive industry planned to establish an automotive information-sharing

and -analysis center (known as *auto-ISAC*) in the USA to document and counter threats.

- **Regulations**. A few countries have restricted or even discussed prohibiting AVs, because of their effects on the job market (India) or because of safety concerns (China). The European Union and a number of US states have restrictions on testing and use. Automakers must work hand-in-hand with policymakers and the technology sector to craft regulations that protect the public without hindering development and all the benefits AVs offer.

- **Third-party concerns**. As many as 4 million professional drivers, including delivery and heavy truck drivers, bus drivers, and taxi and chauffeur drivers, stand to lose their jobs in a fully-AV world—and that is in the USA alone. The International Transport Forum, an industry think tank, warned that, by 2030, automated trucks could cause a decrease in driving jobs of 50–70% in the USA and Europe.

- **Infrastructure**. Airports; train and bus systems; primary, secondary, and tertiary highways; weather stations; commercial enterprises; public offices; and many others will need to work together to map, install sensors, communicate, and provide continuous, updated data so that AVs can effectively, efficiently, and safely navigate riders to their destinations without interference from or to other vehicles. Automakers must collaborate with all these parties and share information for the good of their own business as well as that of the technology and society as a whole.

- **Technology**. To operate, AVs will require up-to-the-minute maps with accurate geography, landmarks, road configurations, construction sites, traffic conditions, gradients, and more. Manufacturers must work with the private and public sectors to complete these maps and invest in research and development to outsource or devise their own technology solutions.

2.6   Case Study—Daimler AG

(see Appendix 125–127)

2.7   Automotive 4.0: A New, Collaborative Model

Automotive industry incumbents such as Daimler, which introduced a partially self-driving truck (see the case study in this chapter), technology companies, and new automakers have all entered the race to the autonomous finish line. Crossing it successfully will require every automaker, retailer, mechanic, and supplier to fully embrace digital technologies—to become, essentially, not an automaker, supplier, or service provider that uses technology, but a technology company that makes supplies, or services cars and trucks.

A variety of stakeholders will need to work together in partnership, which means that original equipment manufacturers (OEMs) must change the way they do business.

Keeping in mind the array of forces that must come together to complete the transformation, OEMs may need to abandon at least some of their old, competitive ways to collaborate with stakeholders all along the value chain—insurers, software companies, hardware manufacturers, cybersecurity companies, infrastructure designers, land-use planners, policymakers, regulators, and more—to form ecosystems in which all work together in partnership toward the common goal of a fully-automated AV world.

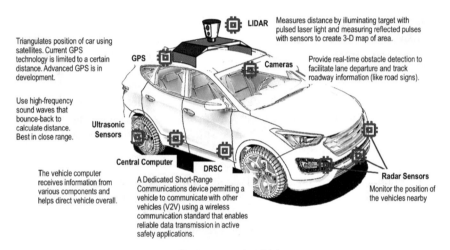

Fully Automated Vehicles

2.8   Glossary of Terms

Autonomous vehicle	A vehicle that uses artificial intelligence in conjunction with other technologies to drive itself with little or no human assistance.
Assembly line	A method of production that moves items along a line of workers who perform a progressive series of repetitive tasks.
Advanced driver-assistance systems (ADAS)	Systems to aid drivers in safely operating vehicles.
Ultrasonic	Sound waves with frequencies undetectable by humans.
Radar	High-frequency electromagnetic waves.

Lidar	Pulses of light, used like radar in sensors.
Gyroscopes	Tools that measure the direction and other aspects of orientation.
Accelerometers	Tools that measure motion and velocity.
Internet of Things (IoT)	Digital technologies that connect humans to inanimate things and things to one another.
Cloud computing	The delivery of computing services over the Internet.
Artificial intelligence	A technology that analyzes and processes data to make decisions with, or in lieu of, humans.
Vision Zero	The goal of eliminating all traffic deaths.
Ride-sharing	The sharing of a vehicle as it travels from one place to another.
Platooning	The grouping of vehicles such as trucks together in a coordinated traffic stream.
On-demand services	The use of a vehicle on demand for specific tasks, including ride-sharing.
Last mile	The gap between where public transportation systems end and the rider's home.
White-hat hacker	A "good" hacker who penetrates a system or network with its owner's permission, usually to find security vulnerabilities tools that measure the direction and other aspects of orientation.

2.9 Questions

1. Name five sectors of the automotive industry that will need to change their business models to remain competitive during the fourth industrial revolution. What changes do you think each of them will need to make, and why?
2. Summarize each of the four industrial revolutions as they relate to the automotive industry. How does Industry 4.0 stand to differ from its predecessors? How might it be the same?
3. What are some ways in which autonomous vehicles might benefit society? List four and discuss how they might make life better.
4. What are some ways in which AVs might be harmful? What might be done to mitigate these potentially negative effects?
5. What challenges and obstacles must the automotive industry overcome to incorporate AV technology into?
6. Their organizations successfully? What challenges must be met to incorporate AVs into the world at large?

Chapter 3
Consumer Revolution

3.1 Consumer and Internet of Things (IoT)

The fourth industrial revolution is affecting not just the workplace, but also personal space. Consumers are increasingly using connected products and devices in their everyday lives, commanding their objects with a word or gesture—just as the inventors of folktales such as *A Thousand and One Nights* imagined long ago.

Already we can perform many tasks by speaking to a *virtual assistant*, a voice-activated device that performs tasks using software. Virtual assistants can check the weather, answer a question, order pizza, and perform a plethora of other functions, usually using the Internet. These devices plus *smart objects*, or objects with software that allows them to be used and controlled remotely, and sensors make up the consumer *Internet of Things* (IoT).

As more and more people use "smart" technologies to help with personal tasks, manufacturers and suppliers of consumer goods are finding that they must consider revising their products, services, and business models to meet the new demands of "connected" consumers.

3.2 The Four Industrial Revolutions: Consumer Goods

1. **Mechanization (eighteenth century)**: The invention of the steam engine enabled machines to assist in the manufacturing process, which enabled more goods to go to market more quickly and cheaply than when they were made completely by hand.
2. **Industrialization: (early twentieth century)**: The development of the assembly line paired with automation brought about *mass production*, in which standardized goods are made in large quantities, making many items widely available and affordable for the first time.

3. **Technology (mid-to-late twentieth century)**: Computer technology and its sibling, globalization, created a vast, worldwide market from which almost anything could be purchased by anyone at any time.
4. **The connected age**: Connected objects not only fulfill consumers' desires for more goods (by searching for them and placing orders) but provide services, as well, acting in concert with other objects to cater to the whims of people and even anticipate their needs.

3.3   Consumer Goods Today

Home appliances and other personal items have largely been inert, operated by hand. The user turns on the object using a manually operated switch, except for those that operate continuously (such as refrigerators). Likewise, when shoppers want to buy an item, they either travel to a store to shop for and buy it, place their purchase over the Internet using a computing device, or call the store to place an order.

The items we use have not, for the most part, communicated with one another or with any outside entity. If they do communicate, they largely convey status reports, as with an oven's signaling that preheating is complete. These "non-smart" items do not collect data or learn, and they are not capable of making decisions.

As a result, users must operate their objects manually each time they are needed—touching a switch to turn on or off a light, pressing buttons to operate kitchen appliances, turning locks to secure doors. If they forget to turn something off, they must return to the item to do so, ask someone else to do it, or wait until later, when they can manually press the switch.

3.4   The Fourth Industrial Revolution: Consumer IoT

The Internet of Things is already changing the way consumers live, transforming their dwellings into "smart homes" whose components will, someday, all work in tandem so that the home is self-operating—in much the way that robots are expected to run factories in the connected age.

For IoT to work, three components must be in place:

- **Physical components**: mechanical and electrical parts,
- **Smart components**: sensors, microprocessors, data storage, controls, software, and, usually, an *operating system*, software that supports a computer's basic functions, and
- **Connectivity components**: ports, antennae, and *protocols*, which are rules governing the exchange or transmission of data between devices, used to connect the product digitally with something else.

Connectivity takes three forms, which can be present together:

- **One-to-one**: An individual product, such as an automobile, connects to a single entity, such as a diagnostic computer.
- **One-to-many**: A central system, such as a virtual assistant, connects to many products at once, such as the alarm, lights, coffee maker, thermostat, and radio to start the day.
- **Many-to-many**: Multiple products connect to many other types of products and often also to external data sources, such as items operating in tandem in a home without a virtual assistant.

Here's how the "smart home" might work:

The consumer's virtual assistant, responding to programmed commands, sounds the morning alarm and turns on lights, adjusts the thermostat, starts the coffee, turns on the news or music, and more. It notifies the autonomous vehicle when it is time to leave for the office, starting it remotely and instructing it to move it out of the garage, so it is warm and ready when it is time to go.

Throughout the day, the home cleans itself, orders pantry and refrigerator items for delivery, monitors and records visitors, and adjusts heating and cooling according to the ambient temperature. The lawn waters itself or not, depending on the climate. Robots may even prepare the evening meal.

When the car starts toward home, the thermostat adjusts, so the home is a comfortable temperature when the consumer arrives. When the car pulls into the drive, the lights turn on, and the door unlocks. Will a robot greet the consumer with a refreshing beverage?

3.5 Going Digital: The Challenges

Timing is everything, and the customer is always right. These truisms are especially relevant in the connected age, when change happens in the blink of an eye and continually, and consumers have become so accustomed to adopting new technologies as soon as they emerge that they expect the businesses they patronize to do the same.

- **Pacing**: Manufacturers will be challenged to keep just the right pace with demands for connected things. If they move too slowly, they will lose customers to competitors with greater *business agility*, defined as the ability to adapt quickly to changes in the market and to their customers' demands. On the other hand, if they move too quickly, they may find themselves with an inventory of products that no one wants.
- **Pricing**: The number of things that could conceivably become "connected" is mind-boggling. Do people want toothbrushes collecting data on how often they brush, and for how long? Not everything that can be digitally connected should be. Adding sensors, Bluetooth and WIFI capabilities, and other tech features cost money, and may raise the price of those toothbrushes—and cause consumers to shun them if they decide the "wired" features aren't necessary.
- **Privacy**: Governments and consumers alike are wary of personal data being collected, processed, and stored, but the machines themselves may rely on that data to learn and improve. For instance, a virtual assistant may alert the home's thermostat that the user has gone to bed, but without data, the thermostat will not know whether to adjust itself or what temperature the user prefers at night.
- **Security**: The more items connected to the Internet, the more likely the home's systems and networks are to be hacked. Hackers have already commandeered security cameras and television sets as well as "smart" nursery monitors. Until consumers can be convinced that the Internet of Things is truly secure, they are not likely to invite additional vulnerabilities to their homes.
- **Customer Service**: With gadgets and appliances designed to work hand-in-hand with other gadgets and appliances, how will customer service representatives help consumers troubleshoot when something goes wrong? If the smart doorbell can't connect with the virtual assistant, which manufacturer should the consumer call?

The retailer who is inadequately prepared to help the customer may lose that person's business.

For all these challenges, however, the consumer Internet of Things is a steadily growing market, with more and more competitors crowding the field. Businesses that can address these issues and stay ahead of the pack—but not too far ahead—by innovating stand a much better chance of success in the new, connected landscape.

3.6 Consumer 4.0: A New, More Agile Model

Business and operating models that make way for a digital future are essential to success in the connected age. Before adapting their products for incorporation into the Internet of Things, business leaders must ask:

Does our business have the technical expertise to make such a change? If not, how will we get it, and at what cost?

How will we implement the changes, and at what pace? How can we structure our teams to respond more quickly and flexibly to shifting consumer demands?

Does connecting my product meet a need or solve a problem? If not, the business might not need to change what it is doing. On the other hand, if management is committed to capturing a share of IoT business, it may be time to put on the thinking cap and innovate.

Do my customers want this? If they don't already know, managers may choose to survey their customers before making the technological leap.

Can the business afford it/Is it worth it? Weighing the costs versus the potential for profit is an important step.

Should we outsource some of the work? Outsourcing can fill talent needs, especially where technology and cybersecurity are concerned. A business has less control over the product and costs with contracted employees, however. On the other hand, hiring employees to do the work in-house is a major commitment—and expense.

Can we offer the same quality of customer service? The customer's questions, problems, and concerns will be radically different from those the business has handled before. How will the business deal with questions involving technology? Setting up a separate staff with technical know-how to handle these issues may be one possibility.

3.7 Case Study—Alibaba

(see Appendix 129–130)

3.8   Amazon Alexa

When Amazon introduced its Alexa-enabled Echo voice assistant in 2015 in the USA, the device was the first of its kind—and met with derision, with some calling it useless and others decrying it as an invasion of privacy. Within just a few years, however, Alexa would be incorporated into more than 20 million devices in more than 80 countries.

Amazon's vision of a smart speaker has grown and changed since the Echo's incipience. Today, the device does so much more than play music; it also orders pizza, helps with online shopping, calls for a ride, tells users about their day, answers questions, makes dinner reservations, turns lights on and off, and so much more—in part because Amazon has made its voice-assisted technology available to other developers to incorporate into their products. Alexa's popularity led to a worldwide boom in voice-enabled virtual assistants; smart speaker sales worldwide were predicted to reach more than $40 billion by 2024.

3.9   The Technologies

Creating a device that operates home appliances, surfs the Internet, integrates with other computing devices, and more required a number of technologies, including:

- **WIFI**: Echo connects to the Internet via its user's WIFI network.
- **Microphones**: A seven-microphone array uses *beamforming technology*, which enables the transmission of a radio signal in a specific direction, and noise cancelation to "hear" the user's voice even from across a room. Echo remains dormant until it hears the activating "hot word" or "wake word"—such as "Alexa"—that activates it to listen for commands or questions.
- **Bluetooth**: Echo can connect to other appliances and devices in the home via Bluetooth.
- **Cloud technology**: The device sends voice commands to a natural voice recognition service in the cloud called Alexa Voice Service, which interprets the commands and sends back the appropriate response.
- **Artificial intelligence**: The driving technology, AI, especially conversational AI, analyzes and responds to requests in near-real-time, and enables Alexa to converse with users.

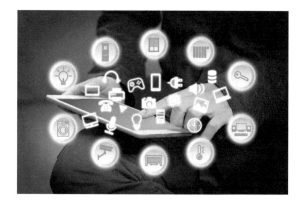

Throughout its continued development of Echo and Alexa, Amazon has remained customer-centric, engaging with consumers to determine their wants and needs, and adjusting its designs accordingly.

When more than 40% of early Echo testers said they would mainly use the speaker for music, for instance, developers enlarged the device from hockey-puck-sized to one that holds a more powerful speaker.

However, Amazon founder, chairman, and CEO Jeff Bezos pressed his vision for the device to be more than a music player, encouraging Amazon's Lab126, which created Alexa, to develop the voice-activated assistant for other tasks, including shopping. Today, virtual assistants can open and display items for sale on users' screens and even make purchases. Echo and other virtual assistants also serve as a central hub from which users can operate smart-home appliances—another innovation that originated in Lab 126.

3.10   The Benefits

Echo has enabled Amazon to not only remain relevant but to position itself as a true innovator in the digital universe. Considered a novelty item first, Alexa may make Amazon as synonymous with consumer computer and Internet use as Google has been for searching and Microsoft Word for word processing.

Other benefits include:

- **Convenience for consumers**. Being able to get information, listen to music, watch videos, operate their homes, make calls, and perform many other tasks with a few words gives customers what they crave: instant gratification with the least effort.
- **Assistance for people with disabilities**. Voice-activated features empower people with disabilities to perform tasks they had previously found difficult or impossible, such as reading a book or programming a thermostat.
- **Sales opportunities for Amazon**. Subscribers to Amazon's Prime premium service can use Echo to shop for groceries from Amazon's supermarket, Whole Foods Market, and have the groceries delivered to them. Amazon offers its

Amazon Music Unlimited Service for Echo by paid subscription. In shopping queries, Echo finds products sold on Amazon's Web site first.

- **Data collection**. Every command or request provides Amazon with information about Echo users, which enables the company to better tailor its services, products, and advertising to individual customers.
- **Partnership opportunities**. The company has teamed with hotels to place Echo speakers in its rooms, allowing customers to continue using Alexa while on the road. Amazon also allows other hardware makers to integrate Alexa into their products, further expanding the company's reach into more homes.

3.11   The Challenges and Lessons Learned

Latency, or lag time between the user's command and the device's response, was one of the first challenges in the development of Alexa. Storing all the data needed to understand and process requests takes one kind of memory; analyzing and responding quickly to those requests take another kind. To bring the experience as close as possible to human-to-human interaction, developers strived to minimize latency. They tried different kinds of processors until finally reaching a latency of 1.5 s— unheard of at the time.

Perhaps even more daunting a challenge was perfecting Alexa's responses so that she sounded human. Conversational AI is complex and difficult technology. Humans don't talk in a linear, logical fashion. Instead, we jump around in our conversations, circle back, and refer to things said before. Context is key to getting it right.

To improve Alexa's comprehension and responses, its home screen displays text and graphics cards showing the user's recent interactions and provides an opportunity to give feedback, so its systems can learn.

To work on speech recognition capabilities, Amazon hired people who had worked at a speech recognition company and also bought two voice-response startup companies. The company conducted numerous tests over several years to gauge which types of responses worked best. The ultimate goal is a natural, intelligent interaction that feels like speaking with a real person—one who knows you as well as you know yourself.

3.12   Glossary of Terms

Virtual assistant	A voice-activated device that performs tasks using software.
Smart objects	Objects containing computers and software that allow them to be used and controlled remotely.
Internet of Things	The network of smart objects transmitting and receiving data, usually over the Internet.

Mass production	A method of production, enabled by the development of the assembly line, in which standardized goods are manufactured in large quantities, making them widely available and affordable.
Operating system	Software that supports a computer's basic functions.
Protocols	Rules governing the exchange or transmission of data between devices.
Business agility	The ability to adapt quickly to changes in the market and shifting customer demands.
Beamformingtechnology	Transmitting a radio signal in a specific direction.

3.13 Questions

1. To function, the "Internet of Things" requires several technological components to work together. Can you name three?
2. List the three types of IoT connectivity and describe each.
3. How does the consumer Internet of Things affect people's daily lives today? Name some examples from your experiences and observations. How do you think the IoT will work in five years? Ten?
4. A number of obstacles exist to IoT's becoming mainstream. Describe three and discuss how you think each might be resolved.
5. Given the growth in consumer demand for IoT-enabled devices, do you think all businesses will eventually need to produce them? What questions should a business consider when deciding whether to do so?

Chapter 4
Energy Revolution

4.1 Energy and "Smart" Grid

How we receive, use, and produce electrical power is changing dramatically. Digital "smart" technologies are enabling utilities to deliver electricity more efficiently, nimbly, and reliably to customers, empowering them in every sense of the word.

For nearly a century, utility customers have received electricity at their homes and businesses and paid a monthly charge to a company that had no competitors. Digital technologies are changing that paradigm, replacing the one-way transmission line with a multi-directional grid, and transforming the passive energy consumer into an active *prosumer*—a combination consumer and producer who chooses how, when, and from whom to purchase power and can even produce electricity to sell to others.

For the power industry, these changes mean new capabilities—and also new challenges.

- Connected "smart" meters that read themselves can:
  - Save money and time,
  - Reduce waste,
  - Improve communication,
  - Increase sustainability.

- Digital transmission and distribution grids can:
  - More easily incorporate energy from diverse sources including renewables such as solar and wind,
  - Stabilize energy flows and adjust them according to demand,
  - Deliver energy efficiently to connected appliances in the Internet of Things,
  - Heal themselves when a power disruption occurs,
  - Re-route transmission when needed to ensure that the power is always on, and more.

But adopting smart technology requires a new, "fully digital" business model in which customers are no longer passive consumers but are energy collaborators and even competitors.

At the same time, connecting to the *smart grid*, in which all phases of the power journey from generation to consumption are digitally interlinked and communicating, opens the electricity infrastructure to outside attacks as never before. The new digital business model will need to incorporate hypervigilant cybersecurity defenses to avoid massive shutdowns.

4.2 The Four Industrial Revolutions: Energy

Starting with the development of the steam engine, each of the four industrial revolutions has also been an energy revolution.

1. **Mechanization (eighteenth century)**: The use of stationary steam engines in factories began around 1760, enabling machines to perform tasks that had previously been done by hand.
2. **Industrialization: (early twentieth century)**: The first power plants and the beginnings of transmission and distribution grids around the turn of the twentieth century enabled electrification and mass production.

3. **Technology (mid-to-late twentieth century)**: Advances in technology, including IT and electronics such as computers, led to a renewable-energy boom starting in the late 1960s.
4. **The connected age**: Digital technologies linking "smart" meters and digitally interconnected transmission and distribution grids are disrupting and democratizing energy production and consumption—putting more power, literally, in customers' hands.

4.3 How the Current System Works

The way electricity makes its way from the power plant to the consumer has remained essentially unchanged since Nikolai Tesla invented the transformer in 1888:

- Electricity generated at power plants from coal, oil, gas, water, or some other source streams over transmission lines to an electrical substation, where it mixes with energy transmitted from other sources.
- From the substation, the electricity flows to distribution transformers near residences or businesses, usually along overhead transmission lines. To cover long distances, substations send power at much higher voltages than electrical circuits can use, so these transformers convert it into lower voltages before distributing it for consumption.
- On-premises meters measure use, and human meter readers check these measurements regularly. Utilities bill their customers for power use based on these readings.

The network of power stations, transmission lines, transformers, and substations is known as the *transmission grid*.

Although this system has worked fairly well, it does have limitations. Reliability is one. Because electricity needs open air to travel along transmission lines, these lines are vulnerable to bad weather. A storm can interrupt the power supply.

Also, calculating how much energy to generate and send can be challenging, requiring a constant balancing act. If demand exceeds supply, generation plants and transmission equipment may falter or even shut down. Consumers may experience drops in voltage known as *brownouts*, so named for the dimming of incandescent lights that results, or *blackouts*, complete loss of power.

As the world's population grows, and as more and more computing devices, appliances, and automobiles run on electricity, power use continues to increase. Keeping up with this demand can be difficult and costly. Many utilities have had to build additional power stations and add *peaking power generators* to augment power supply during times of peak power use.

4.4 The Fourth Industrial Revolution: "Smart" Energy

Digital technologies are now transforming the traditional, one-way electric grid into a smart grid, consisting of digital smart meters, sensors, communications towers, and data centers, acting as a multi-directional conveyor of not only power but also communication and information.

Using digital meters that connect to a *home area network* (*HAN*)—including not just the home or business but all connected appliances, electric vehicles, batteries, and on-premises electricity generators such as solar panels and windmills—the smart grid enables utilities to:

- More accurately predict, and more quickly respond to, fluctuations in energy use, a concept known as *demand response*,
- Sense when equipment is wearing out to avoid breakdowns,
- Measure energy use remotely, and bill customers automatically, reducing the need for personnel, fleet vehicles, and motor fuel,
- Instantly turn the power on and off remotely, instead of having to send technicians to perform this task,
- Track when customers use energy, and bill more for consumption during times of peak demand (and less for energy use when demand is low), a concept known as *dynamic pricing*,

- Reduce energy waste by varying the number of volts delivered to individual devices and appliances, providing only the amount needed,
- Enable consumers to view their energy use over time and in real-time and make energy-saving and cost-saving adjustments,
- Easily accommodate and trace energy inputs from individual power generators such as solar panels and windmills,
- Automatically re-route electricity around trouble spots such as a downed line, reducing the frequency, and duration of power interruptions; and
- Detect when a problem occurs in the line of transmission and make repairs automatically, a concept known as *self-healing*,
- Enable customers to buy and sell power to their neighbors (*peer-to-peer sharing*), increasing efficiency, lowering costs, and conserving energy.

4.5 Going Digital: The Challenges

- **Seamless integration**: Incorporating *Advanced Metering Infrastructure (AMI)*, the integrated system of smart meters, communications networks, and data management systems that make up the smart grid into existing power grids without interrupting energy supply can be complicated, and the technology and labor costly. Smaller utilities, in particular, may find it difficult to pay the up-front costs, even if it means they will save money over the long term.
- **Keeping pace**: Adding to the uncertainty is technology's ever-accelerating pace of change. Traditional, or *legacy*, system components have an expected lifespan of forty years or more, while advances in digital technologies call for replacing AMI components every three to five years or, as advances occur more rapidly, even more often.
- **Interoperability**: Because there are no uniform interoperability standards enabling all smart grid devices to communicate with one another, components from one provider may not work with those from another. This lack of uniformity can make it feel risky to choose an AMI vendor. What if the company goes out of business?
- **Regulations**: Designed for monopolistic systems in which one large utility provides power to every consumer in an area, today's regulatory landscape may not support competition or innovation, and the smart grid depends on both. Current regulations also may not adequately address issues specific to the smart grid such as privacy and security—themselves topics of concern to consumers who worry about their data and user privacy, and to governments worried about hackers' shutting down the power supply.

For all the challenges, countries and cities around the world—in Asia, the Americas, Europe, the Middle East, Africa, and Australia—are deploying smart grid technologies.

4.5.1 Energy 4.0: A New, More Communicative—and Competitive—Model

To adapt to the digital age, utility providers will need to adjust their business models, looking ahead to a paradigm in which multiple energy producers provide power to many people.

To encourage efficient power use, which can help defray some of the costs of going digital, they can use the smart grid to communicate with and educate their customers.

And because digital communications go both ways, they will be able to gather feedback and improve customer satisfaction, critical in an age when consumers can buy their electricity elsewhere or even produce it themselves.

In the future, all utilities will be fully digital. That means that power companies of today, to survive, will have to undergo a digital transformation of their grid, their organization, and their business. They will need to incorporate alternative technologies such as solar and wind power into their portfolios. And they will have to devote more efforts to educating, interacting with, and satisfying their customers who, in the digital age, will literally and figuratively claim more power for themselves.

4.6 Case Study—San Diego Gas and Electric

(see Appendix 131–133)

4.7 Glossary of Terms

Prosumer	A combination consumer and producer who chooses how, when, and from whom to purchase power and can even produce electricity to sell to others.
Smart grid	An electrical grid which all participants in the power journey from generation to consumption are digitally interlinked and communicating.
Transmission grid	The network of power stations, transmission lines, transformers, and substations.
Brownout	The dimming of incandescent lights that can result when energy demand exceeds the amount available.
Blackout	A complete loss of power.
Peaking power generator	Used by power companies to augment electricity supply during times when demand for power is highest.
Home Area Network (HAN)	A network consuming and producing power that includes not just the home or business but all its connected appliances, electric vehicles, batteries, and on-premises electricity generators such as solar panels and windmills.
Demand response	A power provider's ability to predict and respond to fluctuations in energy use.
Dynamic pricing	Tracking when customers use energy, and billing more for consumption during times of peak demand and less for energy use when demand is low.
Self-healing	An energy grid's ability to detect when a problem occurs in the line of transmission and make repairs automatically, without human intervention.
Peer-to-peer sharing	Power consumers' ability to buy and sell power to their neighbors.
Advanced Metering Infrastructure (AMI)	The integrated system of smart meters, communications networks, and data management systems that makes up the smart grid.

4.8 Questions

1. The smart power grid and smart meters enable power companies to do several things they couldn't do before.
2. Name four benefits of digital, "connected" energy transmission.
3. List the three steps from electricity generation to delivery, in place and unchanged since Nikolai Tesla invented the transformer in 1888.
4. What are some limitations of the above, "analog" system? How might the connected smart grid resolve these issues?
5. What challenges exist to utilities' adoption of smart grid technologies?
6. List four of the technologies contained in the smart energy grid used by San Diego Gas & Electric.
7. List four benefits SDG&E has realized since implementing smart grid technologies.
8. What are the three challenges SDG&E faced when going digital, and how did the company meet those challenges?

Chapter 5
Environment Revolution

5.1 Environment and Smart Cities

We are becoming a world of city dwellers. The world's population is not only growing at a more rapid pace than ever before, it is also moving *en masse* to urban centers. While in 2011, half of the 7.6 billion people inhabiting the planet lived in cities, by 2050, as many as 80% of Earth's expected 10 billion population will likely be city dwellers.

Population growth brings a greater demand for the planet's natural resources. Everything we use and consume comes from nature, which is finite: There is only so much to go around.

By-products of human consumption include waste, water and air pollution, and *greenhouse gases* such as carbon monoxide and methane that contribute to *climate change*, human-caused gradual warming of the Earth's atmosphere resulting in extreme weather such as drought, severe storms, and heatwaves, and threatening life forms. These stressors also increase as the number of people rises.

Recognizing the strain on Earth's environment, environmental advocacy groups, governments, private businesses, and individuals are promoting *sustainability*, which is defined as "development that meets the needs of the present without compromising the ability of future generations to meet their own needs, particularly with regard to use and waste of natural resources."

As the planet becomes more urban, "smart," digitally connected cities will become increasingly central to sustainability efforts worldwide.

Using embedded sensors, devices, and wireless communications, smart city technologies collect, share, and analyze data from homes, businesses, transportation systems (including roadways, power and water systems, wastewater systems, weather stations, and more) to help cities function more efficiently and reduce waste and pollution, including greenhouse gas emissions.

© The Editor(s) (if applicable) and The Author(s), under exclusive license
to Springer Nature Singapore Pte Ltd. 2020
J. R. Reagan and M. Singh, *Management 4.0*, Blockchain Technologies,
https://doi.org/10.1007/978-981-15-6751-3_5

Smart city initiatives throughout the globe are testing technology's ability to solve the very problems that, over the centuries, technologies created, beginning with the first industrial revolution.

5.2 The Four Industrial Revolutions: Environment

The four industrial revolutions have wrought cumulative changes in the Earth's environment, caused in part by industrialization and its by-products, and partly by population growth and the resulting rise in consumption of natural resources.

1. **Mechanization (eighteenth century)**: The first industrial revolution began around 1760 in England with the use of human-powered textile weaving machines that soon were powered by water and steam. Thusly began the use of machines to accomplish what previously humans had done by hand. Innovations followed that improved the standard of living, health, and longevity. Lower infant mortality rates and an increase in leisure time combined with these changes brought about a burst in population, from 700 million people to, by 1800, 1 billion.

2. **Industrialization: (early twentieth century)**: By the mid-1800s, the combustion engine, innovations in agriculture, and the invention of the assembly line spurred fossil-fueled extraction and use, which polluted the air and water and started a shift from agricultural, or *agrarian*, societies to urbanization. Further improvements in the standard of living brought about a population explosion—to two billion by 1927 and 4 billion by 1974.

3. **Technology (mid-to-late twentieth century)**, This included the development of the Internet, which improved the economies of developing countries beginning in the mid-1970s, resulting in an even faster increase in the world's population—to 7.6 billion today.

4. **The connected age**: Technologies connecting people and things and collecting data promise ways to counter the negative effects of population growth. Digital technologies help us to monitor and adjust our resource consumption, use renewable energy sources such as wind and solar to produce non-polluting power, and build entire cities aimed at increasing the efficiency of natural resource use and reducing pollution and waste.

5.3 How Cities Work Today

Although governed by a central, municipal body, today's cities operate as discrete organizations, each in charge of a specific service or services. For example, one department or agency might be in charge of providing sewer services, another provides water, and another maintains the city's streets. These providers may be public, administered and funded using taxpayer dollars, or private, run by business. Usually, each provider operates independently of the others and does not share billing, use, or other information with other providers or coordinate services in any way.

Even where residents are digitally connected to their homes and able to monitor and adjust their use of services—turning their thermostats up or down using their phones, for instance, or viewing visitors to their front door via an app that connects to a security camera—these appliances and applications do not share the data they collect with one another nor do they collaborate on service provision.

Development in cities is designed and approved by urban planners with specific parameters in mind, such as the capacity of infrastructures such as streets, public transportation, emergency services, and schools. Often, developers determine characteristics of the structures they build based on market demand, or what people will buy or lease.

As a result of these uncoordinated, even haphazard approaches to development and service provision, cities tend to be crowded, noisy, dirty, and expensive. Public transportation may be unreliable and inconvenient. Crime and safety are concerns, as law enforcement relies on patrols and citizen reports to respond to incidents rather than working to prevent crimes from occurring. Air quality suffers due to auto exhaust and dust generated by traffic. Perhaps these are some reasons why city dwellers are more likely to suffer from chronic stress, according to studies.

5.4 The Digital Revolution

In the city of the future, nearly everything is digitally connected. Embedded sensors and devices citywide—indoors and out—collect information and share it over a centrally operated communications network. This data collection and sharing enable cities to manage their services more effectively, and so help customers use natural resources more wisely.

Smart cities comprise six sectors that must work together to make life more sustainable, efficient, and livable:

- **Smart energy**. Connected home appliances communicate with smart electricity grids, which deliver only the power each appliance needs, recording use in real-time to help power companies manage energy supply and to help consumers manage their use.
- **Smart water**. Digital meters measure and report water consumption and supply, and various technologies help to achieve *zero footprint* water consumption, meaning no new natural resources are used. Examples include rainwater collection and water recycling systems. Sensors and meters can also detect leaks, reducing water waste.
- **Smart mobility**. Sensors detect traffic and road conditions; autonomous vehicles transmit data to help manage traffic flow and reduce congestion and air pollution, as well as help people reach their destinations more efficiently. Sensors in parking spaces can detect which spots are available, enabling drivers and autonomous vehicles to find them more quickly, which reduces traffic.

- **Smart buildings**. *Green buildings*, constructed of environmentally friendly materials, especially recycled materials, now have smart technologies to augment sustainability. Sensors and the Internet of Things, detecting and connecting the systems that provide such services as water, lighting, and heat and air conditioning, help manage those services for maximum efficiency.
- **Smart public services**. Garbage cans and recycling bins equipped with sensors can detect when they are full or empty, helping to improve the efficiency of waste collection. Sensors in light poles detect motion and switch on only when someone is approaching. Sensors also can detect air, noise, and river pollution.
- **Smart integration**. Data is at the heart of what makes a city smart. Data integration centers process and analyze data from sensors, meters, connected things, buildings, infrastructures, drones, and more to synthesize and add efficiency to the delivery of services.

Rather than a disparate set of independent parts, the smart city operates as an interconnected whole, a complex system of systems continuously communicating, all working together. Public health and safety, emergency services, street lighting, housing availability, business operations, and more feed information to software that processes it for use by a variety of public and private entities. This serves to improve service delivery and quality of life. In the process, smart cities, by increasing efficiencies, decrease waste—contributing to a more sustainable environment.

5.5 Smart Cities: Challenges and Obstacles

As utopic as they might sound, smart city development faces a number of obstacles.

- **Sensors**. Although the price of these devices is falling, deploying all the sensors needed to create a world of smart cities—more than 1 trillion, by some estimates—seems formidable, if not impossible.
- **Power**. How will those sensors be powered? Batteries would be impractical since they need replacing; wiring each to the electric grid seems unfeasible, as well.
- **Security**. Every digital connection could possibly be breached, and, connected to everything else in a city, could serve as an entry point for systemwide viruses, data theft, shutdowns, or worse.
- **Cost**. How will existing buildings and infrastructures accommodate the changes needed for the real-time responsiveness that smart cities need?
- **Data**. Until data collection and sharing become automatic, systems will rely on humans to share information about road conditions, emergency situations, traffic jams, and the like. How to motivate people to do so amid privacy concerns?
- **Sustainability**. How will smart cities sustain themselves and remain viable amid the rapid pace of technological change?

All these challenges most likely will be resolved with technologies, current and future ones, either by retrofitting existing grids and structures with smart technologies or building smart cities from the ground up, as in India, which launched an initiative in 2014 to build 100 cities from scratch.

To meet these challenges, city officials—including urban planners—must take the lead, adopting new business models for the cities they manage. They must begin thinking now about how to digitize every service and facility they oversee. Like businesses in every sector, cities need to devise a digital strategy that not only looks to the future but begins today.

Business models suggested for smart city transformation include:

- **Build Own Operate (BOO)**—The city builds the smart city infrastructure and delivers its services, controlling all operations and maintenance.
- **Build Operate Transfer (BOT)**—A city-appointed private partner builds the smart city infrastructure and delivers its services for a specified period, after which the city government takes over.
- **Open Business Model (OPM)**—Infrastructure development and service delivery are open in a free market to all organizations that qualify under the city government's oversight and regulation.
- **Build Operate Manage (BOM)**—A city-appointed private entity oversees infrastructure development and delivery of services as well as management and operations. The city does not assume these roles at a later date.

In spite of the obstacles and challenges ahead, smart cities offer myriad benefits, including increased convenience and quality of life for citizens, costs savings for governments, expanded business opportunities for private enterprise, and conservation of the Earth's resources for future generations.

And they are inevitable: in the future, all cities will be "smart" as digital connectedness becomes the norm. Governments and businesses will need to prepare themselves for cities' connected future by adopting a "digital-first" business model. And as our planet's population continues to grow and resources grow increasingly scarce, the human race as a whole will need to develop a vision that not only includes but *emphasizes* environmental sustainability.

5.6 Songdo International Business District, South Korea

Rather than try to retrofit new technologies into existing infrastructures, the government of Incheon Free Economic Zone in South Korea built the world's first from-scratch "smart city" on the country's northeastern coast. Songdo International Business District occupies 1500 acres of reclaimed Yellow Sea marshland with a city planned in detail for sustainability, livability, and modes of transportation that don't include cars.

Built on 500 million tons of sand dredged from the seabed and poured into the marshy tidal flats, the city rises gleaming and modern, towered over by the country's tallest skyscraper—the 305-m Northeast Asia Trade Tower—and connected by an intricate network of next-generation technologies.

5.7 The Technologies

What really makes Songdo "smart" is its ubiquitous digital technologies, which pervade every home, workplace, and public space.

These technologies converge in connected home appliances, parking lots, streets, factories, and offices to connect people to one another and their places, and to help plant managers make data-driven decisions about their operations. They include:

- Sensors,
- Cameras,
- The Internet of Things,
- Data analytics,
- Television,
- Computers,
- Artificial intelligence,
- Virtual, artificial, and "mixed" reality.

Sensors built into the city's streets regulate traffic, hundreds of cameras linked to emergency services provide security, building heat, water, and lighting also link to sensors for more efficient use. A data screen in a central control center allows for monitoring of all the city's video feeds, data feeds, satellite maps, cameras, and sensors.

5.8   The Benefits

Livability: In addition to the most modern of human-made amenities, Songdo's designers gave equal billing to nature, fashioning a 101-acre park with running and walking trails, fountains, and lakes in the midst of the steel and glass, providing 16–20 miles of bicycle lanes, and instituting a canal system to link neighborhoods together. Forty percent of the city is dedicated to green spaces.

Sustainability: Not all the city's "green" elements are visible to the eye, however. More than 100 of Songdo's buildings are certified by Leadership in Energy and Environmental Design (LEED), the most widely used green-building rating system in the world, for features that include.

- Rainwater collection,
- Reuse of "gray" water from showers, sinks, and other relatively clean forms of wastewater for irrigation, public toilets, industry, and street-cleaning,
- Garbage disposal via pneumatic tubes that funnel the city's waste to a central processing center, which turns it into energy or recycles it.

Convenience: Conceived as a city whose inhabitants would not need to use automobiles, Songdo has ample public transportation and is near an international airport. Its apartment buildings and business sit within twelve minutes of bus or subway stops. Its retail shops, office buildings, parks, medical facilities, and schools are close to housing, and almost always within walking distance of everything. The city produces one-third fewer greenhouse gases than cities of similar size.

5.9   Challenges and Lessons Learned

Cost: Planning and building the Songdo IBD cost about $40 billion. To raise funds, local and national governments partnered with members of the private sector, including a Korean construction company and American real estate, architecture, planning, and technology firms.

Staying current: Innovations and disruptions in technology mean that city administrators must remain nimble, able to shift tactics, and adopt new technologies quickly.

To meet this challenge, Songdo conducts continual research and crowdsources residents' ideas to ensure that the city's "smart" features don't fall behind. Songdo is quite self-consciously a city that will never be finished, but instead, an eternal laboratory testing developing technologies. Perhaps this is why Songdo, whose name means "Pine Island" in homage to the area's many trees, calls itself "the smartest of the world's smart cities."

5.10 Case Study—Cities in South Korea

(see Appendix 135–141)

5.11 Glossary of Terms

Smart city	A city that uses digital technologies to connect people to services and services to infrastructures for more efficient use of resources and improved sustainability.
Greenhouse gases	Emissions into the atmosphere, such as carbon monoxide, carbon dioxide, and methane that contribute to climate change.
Climate change	A human-caused warming of the Earth's atmosphere resulting in extreme weather such as drought, severe storms, and heatwaves, and threatening life forms.
Sustainability	The ability to meet the needs of the present without compromising the ability of future generations to meet their own needs, particularly with regard to the use and waste of natural resources.
Zero footprint	Using natural resources without detracting from the available supply, such as using recycled water for washing clothes.
Green buildings	Buildings constructed of environmentally friendly materials, especially recycled materials, and possibly using digital technologies to improve their sustainability.

5.12 Questions

1. Describe three challenges to smart city development. How do you think these obstacles might be overcome?
2. What six sectors must be present and functioning together for smart cities to be sustainable and livable?
3. Describe the four models for smart city transformation. Which would you prefer for your city, and why?

4. How are "green" buildings in Songdo promoting sustainability? Give several examples.
5. What are some technologies being used in Songdo? How might each contribute to the twin goals of sustainability and livability in smart cities today?

Chapter 6
Finanace Technology

6.1 Finance and Blockchain

In the digital age, how business gets done is speeding up. Legal, financial, and other transactions that typically take days or weeks to complete are now occurring in hours or even minutes using a technology called *blockchain.*

Blockchain is a unique kind of computer coding in which data gets structured in encrypted records, called "blocks," all linked together. Each block in the chain contains an entry in a public *distributed ledger* that everyone in the network can see. Only those with the encryption key can add to the ledger, however, and entries cannot be deleted or reversed.

Invented as the basis for Bitcoin, which is a *cryptocurrency*, a digital form of money, blockchain has many other uses. Its structure enables *peer-to-peer*, or direct, sales of goods and services and payments—in cryptocurrency—without an intermediary, similar to the way email enables the direct exchange of correspondence

J. R. Reagan and M. Singh, *Management 4.0*, Blockchain Technologies,
https://doi.org/10.1007/978-981-15-6751-3_6

without a postal or courier service needed to collect and distribute letters, documents, photographs, and other mail.

If someone generates power using solar panels, for instance, they may choose to sell that energy directly to a neighbor using blockchain, bypassing the utility company.

If someone wants a ride, they may use blockchain to directly hire and pay a driver without using a ride-sharing company.

In the music industry, blockchain would tell musicians, artists, and writers when they make a sale, then pay them instantly, instead of their publishers and galleries sending them periodic statements and royalties. And they could use the data they see on their copy of the transaction ledger to help them market their work, or even to decide what to create next.

In insurance, blockchain enables application-based *"microinsurance"* policies that offer coverage on demand for specific purposes—increased medical insurance while playing football, or short-term car insurance while parked in a high-crime neighborhood—without an agent as go-between.

The financial sector, which includes insurance, may benefit most from blockchain technology, especially cryptocurrency. Cryptocurrency can be deposited and withdrawn immediately in even the most complex transactions such as international currency exchanges or complicated contractual agreements, all of which could, before blockchain, take hours, days, or even weeks to complete.

The technology's impact stands to extend beyond banking to financial services, including trading, cross-border money transfers, and growth capital. Blockchain-enabled peer-to-peer transactions may reduce or eliminate the need for wire services, investment bankers, crowdfunding sites, and even the stock exchange.

6.2   The Four Industrial Revolutions: Finance

1. **Development of the banking system (mid-eighteenth century)**: Humans may always have traded or bartered, but the earliest recorded instances occurred in 9000 BCE, with Neanderthals using cattle as currency.
   Not until 6000 years later was the first monetary coin used: the Sumerian *shekel* dates to pre-3000 BCE.
   The use of paper money began in China, printed by the Song dynasty in 990 CE. Lending and borrowing, as well, have occurred since ancient times, in cultures around the world. With the industrial revolution came the demand for loans of large sums with which to build and operate factories filled with expensive machinery as well as to pay employees, and capital for investors hoping to turn a profit.
2. **Cross-border money wiring (1872)**: Second-industrial-revolution technology included the *telegraph*, the first instantaneous, long-distance form of communication, pre-dating the telephone. The telegraph used *Morse Code*, a system of "dots" and "dashes" sent over a wire by making and breaking an electrical

connection. These short and long "taps," when deciphered in a codebook, formed letters, which made up words.

In 1872, the US company Western Union began using the telegraph to send not just messages but also money. To "wire" funds, a customer paid a telegraph operator in one office, who "sent" them to an operator in another office, who would then pay the recipient. Revolutionary at the time, this service became essential to business: within five years of its launch, Western Union was wiring millions of dollars per year.

3. **Banking 24/7 (mid-twentieth century)**: A new era in finance arrived with the universal credit card (Visa and MasterCard are examples) allowing purchases from a variety of businesses, especially when banks began issuing cards that could be used anywhere in the world.

   In addition, automated teller machines (ATMs), which distributed cash on demand, and Internet banking enabled consumers around-the-clock access to their financial accounts for the first time in history.

4. **Blockchain (early twenty-first century):** Invented in 2008 as the technology behind Bitcoin, blockchain-enabled instantaneous payments using a universal currency. Since then, it has been used for fast, inexpensive, direct transactions, including contracts, payments, loans, and cross-border money transfers.

6.3 How Finance Works Today

The financial system is often characterized as a slow-moving, inefficient beast delayed by massive amounts of redundant paperwork and rife with unnecessary costs—and reluctant to change.

One example is cross-border money transfers, which can take weeks to complete and require many steps.

When a US company needs to make a payment to a foreign person or company, it makes a request to its bank to send the funds. The bank may send the funds to a *correspondent bank*, also known as an *intermediary bank*, a third-party bank that coordinates international monetary transfers and transaction settlements, which sends the payment to a correspondent bank in the destination country, which then transfers the money to the destination company's account at still another bank.

The transfer, in this case, requires four banks as well as originating and receiving businesses or individuals to participate, provide verification, and complete paperwork. The transfer can take days or even weeks to complete, and the risk of fraud and other economic crimes increases with each transaction. Bribery, corruption, fraud, theft, money laundering, and cybercrime touch all areas of the global financial system, resulting in losses of many millions of dollars.

6.4   The Digital Revolution

Blockchain speeds the transfer of money: Bitcoins can be transferred in minutes, to anywhere, from anywhere, and at any time, and without the need for correspondent banks, wire transfer services, or other third parties to be involved.

Increasingly, blockchain technology is being used for non-Bitcoin purposes, as well, including:

- **Smart contracts**: Digital agreements that define rules and set penalties, just as a traditional contract does, but with "if–then" scenarios coded in so that rules are enforced automatically.
- **Compliance**: Banks and other large institutions must secure users' account information against cybercrime and provide proof that the data is secure. With blockchain-based software, they can automatically create a record of who has gained access to information, and control who is allowed to see that data. These records can make security audits much easier.
- **Stock and bond ownership**: Digital "tokens" allow someone to buy or receive a real object digitally, and either store it, give it to someone else or exchange it for the item it represents—whether it be money, music, wine, or stocks.
- **Stock trading**: Blockchain-based platforms allow private companies to issue and trade shares directly, without an intermediary.
- **Crowdfunding**: Software using blockchain technology enables anyone to solicit and give crowdfunding donations without using a third-party service, and to write their own rules. For instance, one project allows donors to vote on how funds contributed to a project would be spent, with those who *give the most* having the most influence.
- **Regulatory reporting and compliance**: Blockchains can provide transparent, accessible records of financial transactions, and can be coded to authorize and automatically report certain transactions.

- **Clearing and settlement**: In the analog world, "T + 3" is the term used for the time needed to clear and settle a transaction: trade day (T) plus three days. With blockchain, the entire lifecycle of a trade—execution, clearing, and settlement—occurs at the trade stage. Trade *is* settlement, occurring more quickly and, without third-party involvement, more cheaply.
- **Accounting and auditing**: Blockchains are databases with context. Each record builds on the last, and all information about a transaction is visible, from its origins to the present moment. This information can be very helpful for accounting and audits.
- **Reducing crime**: Blockchain technology is virtually bulletproof against fraud. Entire networks of computers share—and can view—each transaction. Every change or addition not only requires a special code, or *key*, but it must be verified by a majority of participants in the network, making it much less likely that an unauthorized entity could make changes.

6.5   Going Digital: The Challenges

Blockchain transactions are faster, more transparent, and more secure than conventional digital technologies. But no technology created and used by humans will be completely free of problems. For blockchain, these include:

Theft and fraud. Smart contracts, for instance, are only as smart as their creators. One venture capital funding organization, DAO, lost $55 million in cryptocurrency in 2016 when an investor (and hacker) found a loophole in its crowdfunding "smart contract," developed on the Ethereum blockchain platform, that allowed them to withdraw of the funds. Although the network managed to recover the currency by hacking the chain—proving, again, that it is not completely "bulletproof"—the hacker's identity was never discovered because of encryption tools.

Regulatory oversight. Also, many countries do not regulate cryptocurrency. Others have a mix of policies, which can cause confusion. Investors who suffer losses because of bankruptcy or crime may have no recourse. In the case of DAO, the $55 million stolen would not have been reimbursed had it not been recovered. Likewise, cryptocurrency transferred from one country to another using blockchain could not be recovered or reimbursed if it were lost or stolen.

Initial coin offerings (ICOs), which is the selling of cryptocurrency on the public market, also have risks. Again, regulatory oversight is scant or nonexistent. Some companies issuing ICOs have gone bankrupt, leaving investors with big losses.

But the issues and challenges associated with blockchain do not seem to be hampering its adoption. Financial institutions and software developers around the world are investing in blockchain-based technologies. Banks, lenders, insurers, and others in the financial sector will be wise to prepare themselves for the changes ahead.

Doing so may be difficult for those in a sector that has been steeped in tradition, stability, and resistance to change. Some may choose to move slowly, seeking a single blockchain-based application to adopt or develop, for instance. Blockchain, which is, after all, simply a new way of structuring and organizing data, has been heralded not as a disruptive technology but a foundational one, likely to change every aspect of the sector: the nature of money; how it is raised; how contracts are drawn up and executed.

The promise of more streamlined and profitable operations has many larger institutions around the world investing in blockchain technologies now. For the technology to work, however, the new standards must apply industry-wide: All players in the financial sector must be using cryptocurrency, smart contracts, and other blockchain-based applications, Those who lag behind may be left behind altogether.

6.6   Case Study—Manulife/John Hancock

(see Appendix 143–145)

The Manulife Financial Corporation, with headquarters in Toronto, Canada, is a 130-year-old insurance and wealth management company with additional offices in the USA (where it operates as John Hancock Financial Services, Inc.) and Asia, and annual revenues of $46.5 billion.

Today, the organization is setting itself apart in the very conservative financial sector with a digital transformation project aimed at providing its twenty-first century customers with up-to-date services and positioning itself as a digital innovation leader. To get there, the enterprise has focused first on people, then on processes, and then, at last, on technologies.

6.7   The Technologies

To increase its digital capabilities, Manulife/John Hancock has partnered with a number of technology companies.

AI: One partner developed an artificial intelligence tool to quickly sift through the vast amounts of data generated to help portfolio managers make better decisions.

Data: Another partner created an application enabling customers to share their fitness data in exchange for insurance discounts. The data helps Manulife/John Hancock better serve its customers by offering incentives and push notifications promoting healthier lifestyles, and enables more targeted, customized products and services to fit individual client needs.

Applications: The company's startup Twine launched an application to help customers and their domestic partners collaborate on financial goals, savings, and investments. Its Retirement Plan Services division also offers an application that shows clients their full financial picture and guides them in investment choices.

The Cloud: Cloud technologies allow employees in the organization's various offices to access information about customers from any of the enterprise's locations.

Blockchain: Manulife's Lab of Forward Thinking (LOFT), an innovation laboratory, was exploring the uses of blockchain, starting with ideas such as smart contracts to bring new wealth management customers on board more easily and efficiently.

Internet of Things: Manulife/John Hancock's Vitality life insurance program uses wearables to provide discounts to customers who monitor their lifestyles and demonstrate good practices.

6.8   The Benefits

Customer satisfaction: Increasing its digital offerings enables Manulife/John Hancock to meet customer demands for instant, always-on access to insurance and financial services on their computing devices, wherever they are, responding to claims much more quickly, and speeding up payments to the insured, as well.

Cost savings: The company has reduced personnel costs by shifting more routine customer service functions such as answering requests for information and policy sales to digital, reserving agents for tasks requiring a human touch as helping clients with a major life event such as a death in the family.

Innovation: New technologies such as AI and blockchain enable new types of coverage and services such as the Twine and Vitality applications, and new, digitally enabled features.

6.8.1   Challenges and Lessons Learned

Like banking, insurance is an industry steeped in tradition and caution. Old business models will not suffice in the digital age. Change is possibly the greatest challenge Manulife/John Hancock faces in its digital transformation, including:

Digital knowledge. Rather than try to train or hire workers with IT and development skills, the company partnered with a global consulting and technology firm to move its operations to the cloud and design new offerings on digital platforms.

Culture. Change in any organization begins at the top. To shift thinking enterprise-wide to a "digital first" mindset, Manulife/John Hancock recruited as its senior vice president a chief marketing officer with more than 20 years' experience and hired a CEO whose vision included a "technology-first" approach.

Nimbleness. To keep up with the fast pace of change in the fourth industrial revolution, Manulife/John Hancock established "innovation teams" bringing together workers from various departments to brainstorm new ideas. They gave these teams free rein to work outside corporate bureaucratic restraints—as though they were startup companies—so their innovations could move more quickly to market.

Manulife also launched an "advanced analytics" group made up of data scientists, data engineers, and strategists from its markets around the world to focus on data analytics and marketing; talent analytics; *underwriting*, or the process of deciding whom to give insurance coverage to, the kind of coverage they should get, and what to charge; and fraud detection.

6.9   Glossary of Terms

Blockchain	A type of digital coding in which data gets structured in encrypted records, called "blocks," all linked together.
Distributed ledger	The public compendium of linked entries in a blockchain, visible to all.
Cryptocurrency	Blockchain-based digital currency.
Peer-to-peer sales	Direct transactions in which goods and services and payments are exchanged using cryptocurrency and bypassing intermediaries such as utility companies and banks.
Telegraph	The first instantaneous, long-distance form of communication, pre-dating the telephone.
Morse Code	A system of "dots" and "dashes" sent over a wire by making and breaking an electrical connection, used to communicate via telegraph.
Correspondent bank	A third-party bank that coordinates international monetary transfers and transaction settlements between two other banks.
Intermediary bank	Another term for "correspondent bank".
Smart contracts	Digital agreements that define rules and set penalties just as a traditional contract does, but with "if–then" scenarios coded in so that rules are enforced automatically.
Initial coin offerings	The selling of cryptocurrencies on the public market.
Underwriting	The process of deciding whom to give insurance coverage to, the kind of coverage they should get, and what to charge.

6.10 Questions

1. Describe how blockchain works. What makes it different from conventional computer coding?
2. What are peer-to-peer sales? What are some of its uses? Name three others that you can imagine.
3. What is one advantage of using blockchain for insurance?
4. Briefly describe the four industrial revolutions in the finance sector.
5. What are some of the problems in finance today that blockchain might resolve? How?
6. Name five uses for blockchain beyond cryptocurrency.
7. What are two major challenges for blockchain adoption by the finance sector? How might those be overcome?
8. What changes is Manulife/John Hancock making to its business model to join the digital age?

Chapter 7
Manufacturing Revolution

7.1 Manufacturing and Robotics

Robots are changing the way we work. Machines with *artificial intelligence*, or a computing device's programmed ability to think, learn, and make decisions, are taking their place at work around the world.

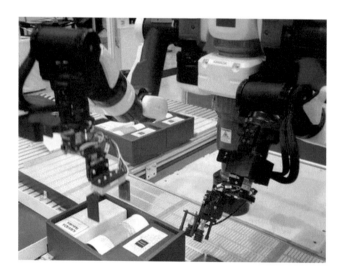

These computer-driven devices can perform repetitive tasks without getting tired, do dangerous jobs without risk of injury, and complete menial chores that free humans to use higher-level skills. They can also sift and sort vast amounts of data to analyze and make assessments in a fraction of the time that humans would need.

J. R. Reagan and M. Singh, *Management 4.0*, Blockchain Technologies, https://doi.org/10.1007/978-981-15-6751-3_7

In popular media, *robots*—machines with artificial intelligence that think, sense, act, and communicate autonomously—have been portrayed chiefly as servants or assistants to humans, performing routine tasks, and helping to solve problems.

In today's world of work, however, robots are filling many roles, as surgical assistants, prescription eyeglass makers, warehouse pallet loaders, underground miners, construction workers, data processors, and more. But the robots are most commonly used for manufacturing work in Industry.

Introduced into factories as long ago as the 1970s, robots were slow to catch on. They were clunky, clumsy, and even a bit hazardous to be around, needing to be bolted to the floor or sequestered by cages so they would not accidentally injure humans.

But robots are becoming increasingly mobile, sensitive, and dexterous, able to accomplish a growing number of feats better, faster, and more inexpensively than humans can do. They can perform repetitive tasks, such as assembly, 24 h a day. They can do potentially hazardous jobs such as welding and heavy lifting without risk of injury. Some even have the fine motor skills to assemble minute, delicate electronics, and micro-electrical-mechanics systems such as those used in automobiles.

In some cases, robots perform their tasks while being directed or manipulated by a human—as in the case of the robot, which operates on a patient under a surgeon's remote guidance. In others, they work independently, doing what they have been programmed to do and, increasingly, improving their performance with time. Others are "collaborative" robots, or *cobots*, working side-by-side with humans and learning from them.

Robots can take many forms. Drones and autonomous vehicles may classify as robots. In factories, they most often perform manual tasks, replacing, or supplementing humans. They may communicate with other robots, as well, soliciting information or sharing knowledge via the cloud and with humans.

7.2 The Four Industrial Revolutions: Manufacturing

To look at the history of manufacturing is to see that machines have become more and more central over time. In the fourth industrial revolution, robots are only the beginning.

Mechanization (eighteenth century): The first industrial revolution saw machines powered by steam and water performs some tasks that humans had done by hand—and do these jobs more quickly and efficiently. Eli Whitney's cotton gin, patented in 1794, separated cotton fiber from seeds fifty times fast than human hands.

Industrialization: (early twentieth century): Electrification and the assembly line, pioneered by automaker Henry Ford in the early twentieth century, enabled mass production, which provided cars and other items to consumers in large numbers and at a much lower cost.

Technology (mid-to-late twentieth century): The introduction of digital technologies such as personal computers and the Internet in the 1990s enabled *automation*, or the use of control systems to operate machinery with little or no human intervention, and globalization.

The connected age: The cloud, the Internet of Things, and robots with artificial intelligence come together to create a *cyber-physical system*, a convergence of the digital and physical in which all things are connected and communicating. This system will allow factories to, essentially, run themselves.

7.3 How Manufacturing Works Today

Although robots are becoming increasingly prevalent, people are still predominant in most factories. Humans may work on the factory floor and also may control robotic arms or other equipment, often using computers, in *assembly lines*, which are a series of workers and machines in a factory by which a succession of identical items is progressively assembled.

Computer-controlled robotic arms, first introduced into factories in the late 1970s, do not the quality or qualities of a robot with artificial intelligence. They have no autonomous capabilities and do not learn or communicate.

Other technologies used in factories include:

Data, or information in digital form, that may help managers improve their systems' efficiencies.

Augmented reality, the use of technology to overlay digital information on an image of something being viewed through a device. These overlays and instructions provided on a set of glasses or other screen help workers do their jobs, learn new skills, and repair and maintain equipment.

7.4   The Fourth Industrial Revolution: "Smart" Factories

In the factory of tomorrow, everything will be connected and communicating via the cloud, powered by artificial intelligence.

Robots connected to factory machines, for instance, will detect when a machine needs maintenance or when a breakdown is imminent. They may be programmed to order parts, supplies, and equipment from suppliers autonomously, and then to make needed repairs.

Connected to customers, they may autonomously adjust production schedules as needed, scaling up or down to accommodate changes in demand.

They may interact with humans, working with them on projects that require a human touch, and providing data and analysis to help inform decisions.

And they will be able to do these things around the clock and at a lower cost than human labor commands. Some predict that robots will enable *reshoring*, the return home of a manufacturing plant from the foreign country where the business has *offshored* it or placed it in a country different from where the home office is located.

Businesses often want to reshore because of shipping costs, government incentives, a shorter supply chain, or customer satisfaction.

7.5   Benefits of Robots

Robots can:

Enable manufacturers to produce their goods locally and to increase productivity while lowering wages.

Improve flexibility and responsiveness to the market, which is needed to compete in today's fast-changing business world.

In the digital age, innovation proceeds at a dizzying pace, shortening products' life spans as new inventions or improvements come along quickly, and requiring frequent changes in what factories produce and how they make them.

Training workers to make something new can be an onerous process. In a cyber-physical system, however, only one robot may need to be programmed and trained. The trained machine can then share its knowledge with the other robots in the plant, training them in real-time via the cloud.

7.6 Going Digital: The Challenges

Equipment and infrastructure costs: Creating a factory that has little or no human presence requires the purchase of sensors, robots, computers, software, and other components, and their maintenance and frequent updates. Affordability may be especially challenging as innovations and improvements quickly render technologies obsolete. And early adopters may see little return on investment at first, unable to take full advantage of the features, a cyber-physical system offers until others, such as suppliers and customers, are connected, too.

Labor costs: Although it's commonly thought that the price of robots will be offset by savings on human labor, personnel costs may not decrease as much as assumed and could even go up. Robots and cobots may replace many manual workers, but these jobs tend to pay a lower wage, anyway. Companies will need to hire more people with skills in programming, analytics, system integration, interaction design, and other related functions—higher-level jobs that command higher pay.

Safety: Robots must learn how to work in the proximity of people, and vice-versa, while avoiding causing injuries to the more-vulnerable humans.

Resistance to change: Humans are naturally resistant to change. And factories *botsourcing*, or replacing humans with robots, can add to the problems. Manufacturers could find themselves dealing not only with public relations challenges as clients and customers resist botsourcing, but also with resentful employees—potentially creating a human resources issue.

Chances are, the transition to fully cyber-physical factories will be gradual, with many of these obstacles resolved in time. Eventually, all factories will be fully automated cyber-physical systems operated completely or at least in part by intelligent and autonomous machines.

To compete in this new landscape, manufacturers will need to:

Develop a digital strategy that considers the technologies needed.
Analyze which jobs stand to be displaced by robots, and consider retraining the employees now in those positions.

Incorporate robots gradually into the workplace, to help employees adapt to working with them.

Collaborate with others on their value chain, working with suppliers and customers to develop new ways of doing business in a shared cyber-physical environment—keeping in mind that the business model of tomorrow supports flexibility, rapid scaling—down or up—and innovation.

7.7 Adidas's 'Speedfactory' Plants

Until a few years ago, the athletic-wear manufacturer Adidas produced its shoes and other sporting gear in factories located primarily in Southeast Asia, then shipped its products to retail outlets around the world—a process that took many months and entailed many challenges.

Producing in cheaper labor markets distant from corporate headquarters ("off-shoring") meant that the company had to split its operations. While its in-house staff designed and developed products and handled marketing and sales, factory laborers on the other side of the globe stitched, glued, and laced Adidas's shoes and other sportswear. The time from design to production for a new style averaged 18 months.

Also, exercising quality control was difficult across the miles, and scaling was a monumental task: sprawling factory complexes only made shoes in lots of 20,000 per shoe size. Distribution was slow, as well, with shipping from factories to retail outlets taking as long as six weeks.

Today, Adidas is using automation and digitization to resolve these and other challenges. "Speedfactories" produced a single, custom-made pair of sneakers in as little as one day after a customer ordered, using only a small human workforce.

Located in Ansbach, Germany, near Adidas's corporate headquarters, the smaller, mostly automated plant used digital designs that could be altered according to even a single consumer's needs and desires and sent to robotic arms and 3D printers for precise custom fabrication. The company opened a second Speedfactory in the USA in 2017.

Because the factories were near their target markets, Speedfactory shipments could arrive at customers' doors soon after they had placed their orders. Like Adidas' shoes, these factories were designed for speed—enabled by connected technologies. The final step still must be taken by humans, though, because machines still cannot: lacing the shoes.

7.8 The Technologies

Almost every step of shoemaking at Speedfactory happens digitally, using:

- **Data**. Digital information goes into the products starting with customer specifications, including foot size, gait, and other qualities as well as design features.
- **Virtual reality (VR)**. The design of the shoes happens on a computer screen, with digital prototypes tested in a VR environment. The manufacturing process also gets simulated using a *digital twin*, or digital replica of an existing environment.
- **Robotics**. Robot machines knit the uppers of the Adidas trainer shoes and work in other aspects of the factory's operations.
- **Artificial intelligence**. Machines in the factory communicate with one another, with customers and clients, with employees and managers, and with suppliers to run the factory with little human intervention.
- **3D printing**. Printers produce the shoes' midsoles to be joined with the knitted uppers and other parts of the shoe.
- **Internet of Things**. Sensors, the cloud, and other technologies monitor and facilitate the communication between machines and people to enable the plant's autonomous operation.

7.9 Benefits

- **Shorter lead time**: Speedfactories are true to their name. These digital-first factories reduced the time from design to finished goods—previously 12 to 18 months to 45 days.
- **Smaller batch size**: Speedfactories can produce just 500 pairs of a shoe at a time as compared to 50,000-to-100,000-pair batches in its larger mass-production plant. The smaller batches mean the company can respond to consumer demands more rapidly and innovate new designs or the use of new materials more frequently.
- **Lower costs**: Using advanced technologies and placing smaller factories nearer the customer base means shipping and transportation costs decrease. Automating manufacturing processes saves on labor costs, as well.

- **Customer satisfaction**. Shoppers are no longer content to wait to wear the styles they see on social media but demand instant satisfaction. Eighteen months between the start of design to production was much too long: by the time, a new shoe hit the market, it was already outdated. Speedfactories enabled people to custom design their own shoes, place the order, and receive the finished pair within as little as a few days.

7.10   Challenges and Lessons Learned

Change in itself was a challenge for Adidas. In business for more than 40 years, the company resisted change at first. Its journey from conventional manufacturing to digital-first Speedfactories was long and gradual, spurred by consumer demand, and competition. Competitors were partnering with technology companies to transform themselves; Adidas opted to hire hundreds of engineers and software engineers. To design and build its first Speedfactory plant, the company partnered with a manufacturer of precision parts and components as well as a robot manufacturer and a technology company.

Nimbleness, mandatory in the "consumer demand" era, was difficult when producing shoes in large batches. The company launched a data initiative in 2014 to analyze trends and make predictions but could not keep up with fast-changing consumer demands. Next, Adidas tried speeding up the supply chain, aiming to reduce wasted time and speed up orders, but could only shave days or weeks off the process. At last, the manufacturer reached its desired level of flexibility and responsiveness with its connected, automated, and customizable Speedfactory model.

Competitiveness. Adidas's competitors were also launching digital initiatives and increasing automation. To stay ahead, the company would need to incorporate connected technologies throughout its supply chain, including logistics and smart warehousing.

Knowledge. As noted above, Adidas hired engineers in-house to help automate its manufacturing processes. It also contracted with technology companies to design and build its Speedfactory plant using a *digital twin*, or digital replica of an existing environment, often used for testing of workflows, traffic, and other relevant traits before a factory is built or while an existing plant is in use.

To truly compete in the connected age, Adidas would need to incorporate smart technologies through its entire operation and supply chain, including smart warehousing and logistics. To do so, the enterprise would need to find partners with special knowledge and skills in digital business transformation.

7.11 Case Study—Adidas Speedfactories

(see Appendix 168–172)

7.12 Glossary of Terms

Artificial intelligence	A computing device's programmed ability to think, learn, and make decisions.
Robots	Machines with artificial intelligence that think, sense, act, and communicate autonomously.
Cobots	"Collaborative" robots that work side-by-side with humans and learn from them.
Assembly lines	In a factory, a series of workers and machines that progressively assemble succession of identical items.
Data	Information in digital form.
Reshoring	The return of an offshore manufacturing plant to its country of origin.
Offshoring	Establishing a manufacturing plant in a different country from where the home office is located.
Botsourcing	Replacing human workers with robots.
Digital twin	A digital replica of an existing environment, often used for testing of workflows, traffic, and other relevant traits before a factory is built or while an existing plant is in use.

7.13 Questions

1. What are some tasks that robots can perform better than humans?
2. What are the chief characteristics of the four industrial revolutions as applied to manufacturing?
3. How is manufacturing changing in the connected age?
4. What business benefits might robots bring to the manufacturing sector?
5. What are three challenges to the use of robots and robotics in manufacturing, and how can plant managers overcome them?
6. Describe the changes in Adidas's manufacturing process brought about by its Speedfactory plants.
7. What more could the company do to truly "go digital"?

Chapter 8
Media and Entertainment Revolution

8.1 Media and Entertainment and Virtual Reality (VR)

Media and entertainment are moving outside the box and beyond the screen. After nearly a century as the predominant medium for entertainment, screens—be it in televisions, movie theaters, computers, or mobile devices—are giving way to digital experiences that seem to surround the viewer, much like reality.

In the fourth industrial revolution, the definition of "media" is changing. News can be something we witness first-hand, asking questions at a press conference, touring a refugee camp, or fleeing the scene of a disaster. Instead of watching television, we participate in its shows, walking through the house that needs remodeling, cooking with the chefs, performing a comedy routine on the stage, or heckling the comic from the audience.

Entertainment, too, is leaping off the screen and into our surroundings, turning us from passive viewers into active participants in events we never thought possible:

We can soar with eagles. We can walk on distant planets. We can travel around the world, and to distant, imagined worlds of our own creations.

Virtual reality (VR), a computer-generated, three-dimensional environment that immerses the user, is changing the face of media, with the help of technologies including:

- *Enhanced optics* such as special glasses that help us see in 3D,
- *Holograms*, or projected, photographic images made of light,
- *Haptic technologies including* gloves and other items of clothing that allow us to "touch" and be "touched,"
- *Biosensing,* which detects body movements,
- *Telepresence*, which brings people together virtually into a shared space,
- Technologies to stimulate the senses of smell and taste, enhancing the *mixed reality* experience, which is the mixing of physical and virtual realities.

With VR, we can attend sports events and concerts, even participating in the game or performance; go spelunking in caves, mountain climbing, or sky diving; and even meet far-off friends in a virtual café, all without leaving home. Or we can indulge in gaming or other activities in public VR "parlors," similar to video gaming arcades, in which participants play solo or with others.

Having previously been passive viewers of scenarios scripted by someone else, we may be interactive participants in situations of our own choosing—and even our own devising—in worlds real and imagined where the sky is *not* the limit.

8.2   The Four Industrial Revolutions: Media and Entertainment

1. **Mechanization (eighteenth century) and the printed word**: Johannes Gutenberg introduces the moveable-type printing press between 1440 and 1450, enabling mass production of written materials for the first time and ushering in the *Modern Age* (until 1945).
2. **Industrialization (early twentieth century) and motion pictures**: In 1895, Auguste and Louis Lumière conduct the first public movie screening on their *cinematographe*, a combination movie camera, film processor, and projector, in Paris, France. In 1927, twenty-one-year-old Philo Taylor Farnsworth demonstrates his new invention, electronic television, in San Francisco, California, USA.
3. **Technology (mid-to-late twentieth century) and the Internet**: The United States Department of Defense's ARPANET, or Advanced Research Projects Agency Network, develops the Internet beginning in 1983. Computer scientist Tim Berners-Lee invents the World Wide Web in 1990, introducing Web sites and hyperlinks to access information on the Internet, which led to its popularity among the public.

4. **The connected age and mixed reality**: Digital technologies, including VR, usher in a new era of media and entertainment, one that transforms the "viewer" of videos, print media, and films and player of screen-based games into an interactive participant in immersive worlds, including ones that they create themselves.
5. **Augmented reality**: While virtual reality makes the user feel they are in a different place via 3D, augmented reality is achieved through "smart glasses" (transparent), allowing you to see everything in front of you, but with a data overlay. You may see your surroundings clearly, but fabricated images or animate objects appear in your field of vision as well. For example, Pokemon Go projects a Pokemon character into your visual field.
6. **XR**: Virtual reality, augmented reality, mixed reality—do we really need another reality? XR refers to the basket in which *all these previous terms fall into*. In essence, XR speaks to the connectivity of all the other "realities," allowing the user to see in 3D, touch via haptics, and experience augmented reality via data overlays which project a virtual world or objects into your visual field.

8.3   How Media and Entertainment Work Today

Since the invention of the printing press, followed by electronic media, humans have consumed mass media in a passive, one-directional manner—reading books and newspapers, watching movies and television, and listening to music. Sporting events involve watching from the stands or on-screens. Games, although interactive, fall short of immersive, since the action takes place on a two-dimensional board or screen.

If we choose to attend concerts and sports events, we need tickets which are limited in supply, must travel to the venues where the events are taking place and may wait in line to enter the venue and claim our seat. To play games, we may wait for access to our preferred table or our ability to participate limited by space availability and demand as well as our own ability to travel.

The Internet and the World Wide Web changed the rules of engagement, enabling, for the first time, interactivity in *real-time*, or in the moment that something is occurring. Online, we can comment on what we read, watch, and hear and read others' comments; play games with people who connect with us online; create and post our own media or others to read, watch, or hear; collaborate on projects, and more. No longer are we solely passive consumers of the media others create and present to us, but commenters, critics, and creators. And yet, with the Internet, the screen remains the mediating space, a "fourth wall" separating us from the world we are watching and hearing.

8.4   The Fourth Industrial Revolution: Smart Media

The concept of VR may have originated with *stereoscopic viewers*, which date as far back as the early nineteenth century. These presented dual images that merged before the viewer's eyes into a scene that appeared as three-dimensional.

In the mid-1950s, an American cinematographer, Morton Heilig, invented the Sensorama, an arcade-style machine that used stereo speakers, a stereoscopic 3D display, fans, smell generators, and a vibrating chair to stimulate all the senses. In 1960, he followed up with the first head-mounted cinematographic display that included sound.

Various attempts to develop commercially viable, immersive VR have followed, but not until 2010 did the technology fully emerge with the invention, in 2010, of the Oculus Rift headset by American teenager Palmer Luckey. Many other developers have joined the race to create VR hardware, apps, and programming that push the boundaries ever farther beyond the screen.

Those technologies include *mixed reality light-field technology* that projects the virtual world onto the real world.

8.5 Benefits of Virtual Reality

VR *live streaming,* which presents online events viewed in real time, places users in a concert, sports, or other event venue without having to travel there, and can even place them on the stage or on the field of play.

Clothing using haptic sensors bring the sense of touch—being touched, touching, and sensing movement—to an experience for more interactivity.

Scents such as guava, coconut, and suntan lotion accompany immersive sights and sounds to transport the user to tropical lands.

And all these experiences may someday become available without the user's having to wear anything: VR without a headset, virtual touch without special clothing, virtual scents transmitted using ultrasound, and more.

8.6 Going Digital: The Challenges

After a long period of incremental progress toward VR, media and entertainment companies now find themselves having to catch up quickly in order not to be left behind. The very nature of storytelling is shifting from a linear narrative created and presented for passive consumption to one that is interactive, which must adapt smoothly to user choices for the most lifelike experience possible.

Fortunately, media and entertainment are ahead of many sectors in adopting new technologies and planning digital strategies. Virtual reality and other immersive technologies, however, will pose special challenges:

- **A shift in paradigms**. Media companies must switch their content and presentation from:

  - Screen-based to immersive,
  - Linear to holistic,
  - Designed for passive consumption to inviting interactive participation,
  - Long-form to short, since VR tends to be more taxing than relaxing.

- **Innovation**. Success will require changes not only in how stories are told but also how they are presented, as well as a focus on innovation in the technologies themselves. Movie theaters may disappear or will need to offer experiences far beyond what audiences can get at home, especially as on-demand access to content becomes the norm.
- **Personalization**. Data analytics will be increasingly important for virtual and mixed realities to deliver individualized, personalized experiences, which users are increasingly coming to expect.
- **Value**. Deriving value from media and entertainment products will pose a challenge. For many years, the industry has been able to support itself with advertisements, which rely on captive audiences to watch them, and ticket sales, which depend on scarcity of resources: To pay a high price for an experience, the

consumer must see it as one that is not only extremely desirable but also limited in availability.

- **Going digital**. As in other industries, digital innovation is key to competitiveness. Media and entertainment businesses will cease to be merely purveyors of content but will seek mergers and acquisitions or other kinds of partnerships with companies that provide the technology itself or distribute the content in new ways.

8.7 Case Study (see Appendix)—Google VR

(see Appendix 173–176)

8.8 Glossary of Terms

Virtual reality (VR)	A computer-generated, three-dimensional environment that immerses the user.
Enhanced optics	3D or other glasses that help the wearer see virtual, augmented, and mixed realities.
Holograms	Projected, photographic images made of light.
Haptic technologies	Gloves and other items of clothing that allow us to "touch" and be "touched".
Biosensing	Technology that detects body movements.
Telepresence	The use of technology to bring people together virtually into a shared space.
Mixed reality	The mixing of physical and virtual realities.
Real-time	The moment that something is occurring.
Stereoscopic viewers	Viewers presenting dual images that merge before the viewer's eyes into a scene that appears as three-dimensional—dating to the nineteenth century.
Mixed reality light-field technology	A type of digital technology that projects the virtual world onto the real world.
VR live streaming	A form of VR that presents events online for viewing in real time, places users in a concert, sports, or other event venue without having to travel there, and can even place them on the stage or, in sports, on the field of play.
Accelerometer	An instrument that detects three-dimensional movement. Such as walking or throwing something.
Gyroscope	An instrument that detects angular movement, such as looking up or down.

Magnetometer	A form of compass that tracks the direction of movement.
Positional audio	The use of multiple speakers to provide the illusion of the VR user's being surrounded.

8.9   Questions

1. Discuss some of the ways that VR is changing media and entertainment. Are these changes good or bad? Why?
2. Name three technologies that help VR to present a more realistic experience.
3. Describe the effects of the four industrial revolutions on media and entertainment.
4. What is the "fourth wall" in media, and how does it take form today?
5. Describe the development of VR over the years. Why do you think it has taken so long to come to fruition?
6. What challenges will the media and entertainment industries face as VR becomes more popular? List three challenges, explaining how the industry can meet those challenges.
7. What are some experiences that Google VR can offer to its users?
8. What are "console wars," and what could Google VR do to win this battle?

Chapter 9
Medical Revolution

9.1 Medicine and Wearables

It has long been said that knowledge is power. And as mobile technologies provide us with more information literally at our fingertips than ever before, people are taking increasing control of their lives, including their health.

As a result, the ages-old "doctor-patient" paradigm is turning on its head. Data from health-related apps and wearables combined with ready access to medical knowledge—including medical records—are changing people's relationships with their healthcare providers. No longer are they passive patients, but informed consumers.

This profound shift empowers the patient, making them a partner in their healthcare decisions, helping them to better choose which provider to see, and even enabling them to avoid the doctor's office or hospital altogether.

The connected consumer can now say "no, thanks" to medications whose side effects they do not like, for instance, or make dietary or behavioral choices that enable them to improve their health without medical treatment at all.

And as apps and wearables offer cheaper and more convenient ways to diagnose and monitor health, get advice, and self-treat, providers may find themselves having to compete for business in ways they had never envisioned.

J. R. Reagan and M. Singh, *Management 4.0*, Blockchain Technologies, https://doi.org/10.1007/978-981-15-6751-3_9

Survival in the connected age will require a profound shift in mindset among medical practitioners and facilities such as clinics and hospitals. With customers increasingly able to care for themselves, providers may find their services in less demand and their relevance at risk. To attract customers and sustain growth, they will need to transform their business models to become customer-centric, aiming not only to improve patient outcomes but also to create value for healthcare consumers.

9.1.1   The Four Industrial Revolutions: Medicine

1. **Anatomy**: The sixteenth–eighteenth centuries brought important discoveries regarding human anatomy and systems, including how human circulation works, the existence of cells, neurotransmitters, the immune system, and human chemistry, including enzymes, hormones, and metabolism.
2. **Epidemiology**: Scientific research yielded information on the causes and treatment of disease in the nineteenth century, including discoveries by Louis Pasteur and Robert Koch regarding microbes and bacteria and their roles in illness. This century saw significant advances in *immunology*, the study of the immune system.
3. **Technology**: At the turn of the twentieth century, Wilhelm Conrad Roentgen discovered X-rays, earning the first Nobel Prize for physics. Ultrasound, *angiography*, which delineates the interior of the heart, CAT scans, and *magnet resonance imagery* (*MRI*), which displays very small structures in the body, also were developed during this century. The end of the century heralded the sequencing of the human genome.
4. **Connection**: Digital technologies, including wearables and apps, enable unprecedented access to medical knowledge, including knowledge of one's own body, empowering people to self-diagnose, self-treat, and even head off illness with *predictive analytics*, the use of data to predict the onset of illness and disease.

9.1.2 How Healthcare Works Today

Today's healthcare system is primarily reactive, with people seeking medical advice and treatment only after symptoms occur. For most of the history of health care, medical practitioners had exclusive access to the latest research and findings, and patients relied solely on their physicians' expertise for diagnosing and prescribing treatments. Providers dictated the course of treatment, including where and when it would occur and for how long. Patients might have no advance information regarding costs and had to pay whatever they were billed, minus amounts covered by insurance.

In-office visits have presented, in most cases, consumers' only opportunities to learn about their underlying physical condition. Providers were the sole possessors of the technology needed to examine and measure vital statistics from blood pressure to heart health and more. To gain access to this equipment and the data it provides, consumers have needed to schedule appointments ahead of time (sometimes, months in advance), traveling to a doctor's office, clinic, or hospital during set hours. Then, they have had to wait on-site until the provider is ready to see them, pay additional fees for each new test, and wait to get results—possibly in a second office visit where the doctors would provide a diagnosis and prescription for treatment.

If no diagnosis is evident, further appointments may be necessary, each incurring not only billed expenses but also the costs of transportation, time off work, and family care.

Office physicians and other medical professionals tend to work when everyone else does. People needing medical care at other times, such as at night or on the weekends, usually must visit special 24-h clinics or hospital emergency rooms where the price of care is much higher.

The current system, in other words, is designed primarily to serve the needs of healthcare providers, not consumers. Patients are at these professionals' mercy, often seeing whichever physician they have been referred to, checking into whichever hospital or clinic they have been assigned to, and having little say regarding their treatment.

9.1.3 The Healthcare Revolution

The connected age is ushering in a new era of consumer expectations in virtually every industry, including health care. Mobile technologies, in particular, are enabling people to access medical advice, information, and even treatments on demand.

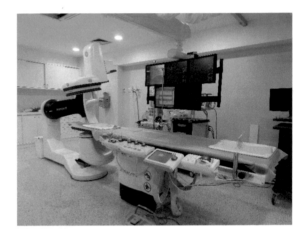

Much of this personal data comes from *wearable technologies*, those containing electronics sensors that we wear, carry, or ingest and that captures and exchanges data with a digital network. Examples include watches, clothing, glasses, mobile phones, and *digital pills*, which contain sensors that, when swallowed, send information to a medical practitioner.

Wearables are tracking and monitoring an increasingly broad range of behaviors and biological traits, including:

- Heart rate,
- Blood pressure,
- Blood glucose levels,
- Heart condition,
- Concussions,
- Ear infections,
- Skin cancer,
- Parkinson's disease,
- Alzheimer's disease,
- Body temperature,
- Sleep patterns,
- Stress levels,
- Weight,
- Activity levels,
- Moods.

The data generated may go directly to a doctor, nurse, or healthcare facility. It may be added to the user's *electronic health record* (*EHR*), which digitally records and displays an individual's health and medical care information and is available to the patient as well as their healthcare providers. Or it may be for the user's eyes only to help track and understand how they might improve their health, or to sound alerts that medical attention may be needed.

No longer, then, does the consumer necessarily have to visit a doctor's office to have vital statistics or symptoms examined for signs of illness or disease. Instead, they can scan and identify many conditions using wearable technologies and apps, deciding for themselves the next course of action.

9.1.4 Connected Health: Challenges

A new paradigm for patients almost certainly means dramatic changes in health-care providers' business models. Previously based on assumptions that practitioners were the experts, and so had full authority over their patients, the medical model of tomorrow will treat the patient as an empowered, even equal, partner. That means, like their customers, they must incorporate digital technologies into every aspect of their practice and business.

The business case for doing so is multi-fold. Using connected digital technologies increases efficiencies, reducing paperwork for the provider while speeding consultations, often increasing the number of patients per day a practitioner can see. For example, special cameras now enable eye exams without first dilating the pupils, speeding patients through their screenings, and taking less of the ophthalmologist's time. Also, by providing individualized information, wearables and their connected apps enable treatments to be more personalized, which promise to improve outcomes.

But obstacles to change abound, including:

- **Responsibility**. Who will process all the added data that connected health sends to providers? Many physicians, especially primary-care doctors, already struggle with their existing workloads. Should they answer emails, schedule tele-visits, and review the data coming in from patients' wearables and apps, as well? Added staff may be needed for these transactions.
- **Payment**. Many insurers will reimburse their clients only for in-office visits, based on the belief that making it easier to consult with one's physician will result in more frivolous appointments. One solution might be "value-based" payments that hold providers accountable for the costs of their patients' care. Such a change would encourage physicians and other providers to offer more efficient modes of care using digital technologies.
- **Regulations**. Government regulators are slow to catch up to the digital space in general. Working slowly to approve new health-related wearables and apps can impede startups from getting funding and bog down innovation. At the same time, a lack of expertise makes moving too fast potentially risky, especially where users' well-being depends on a device's functioning properly and as promised.

Eventually, these and other challenges will work themselves out as connected health becomes the rule rather than the exception.

In the meantime, medical providers will need to adjust their business models from the prevailing top-down, autocratic approach to one that asks, first and foremost, "What does the customer want?" Then, they will have to find ways to meet those expectations digitally in a connected way—not only for their patients' health, but also for their own.

9.2   Case Study—The Medical Wearables Industry

(see Appendix 157–159)

9.3   Glossary of Terms

Immunology	The study of the immune system.
Angiography	The use of X-rays to delineate the interior of the heart.
Magnetic resonance imagery (MRI)	A non-invasive medical imaging technique using magnetic fields and radio waves to display very small structures in the body.
Wearable technologies	Electronic sensors that we wear, carry, or ingest and that capture and exchange data with a digital network.
Medical wearables	Devices worn on the body that collect, monitor, and measure data about the wearer's vital signs and physical health.
Pedometer	An instrument that counts its user's steps.
Internet of medical things	Digital medical devices connected to their users and to one another.
Cloud computing	The delivery of digital services via the Internet.
RFID	Radio-frequency identification.

9.4   Questions

1. How is technology changing the relationships between medical caregivers and their patients? Is this good or bad? Why?
2. How will caregivers need to adapt to these new paradigms to remain competitive?
3. What are the four industrial revolutions in medicine, and how did each transform the sector?
4. Name five human traits wearables can track. Can you think of three more not listed here that might be useful to healthcare providers and consumers?
5. What is considered the first medical wearable, invented by whom, and when?
6. Name three technologies important to the medical Internet of Things.
7. Name three benefits of medical wearables.
8. Name three challenges or obstacles the medical wearables industry is facing. What do you think can be done to resolve or minimize these challenges?

Chapter 10
Retail Revolution

10.1 Retail and Cloud

In the retail game, all the rules are changing. The prevailing model in which shoppers enter a store, purchase items from the shelves or racks with the help of a customer assistant, and carry their purchases out the door is becoming a relic of the past.

Today, most consumers research items on the Internet before buying. Many never visit a store. Most who do shop in a store found the retailer in an online search. And when they go, they will likely have a much different experience than even a few years ago.

Digital technologies including artificial intelligence, the Internet of Things, augmented and virtual reality, and big data analytics are transforming the retail experience for the buyer—one that, thanks to cloud-based technologies, enables the shopper to start and stop multiple times, switching among computing devices and platforms—a virtual assistant, a phone call, an in-person visit to the store—as smoothly as if they'd completed the entire transaction in a single session. At the same time, these technologies provide the merchant with completely new ways to fulfill customer wants and needs.

J. R. Reagan and M. Singh, *Management 4.0*, Blockchain Technologies, https://doi.org/10.1007/978-981-15-6751-3_10

The challenge, for retailers, lies not only in staying in the game but also in getting ahead of it. Doing so means providing customers with what they have come to expect—tailoring the shopping experience uniquely for them, and in the instant—while anticipating upcoming trends and innovating new experiences.

At the same time, retailers must compete on an increasingly vast and diverse playing field against merchants around the world. The digital age provides unprecedented opportunities for even the smallest mom-and-pop shops to not only better understand and serve their customers, but also sell their goods and services on a global scale.

10.1.1   The Four Industrial Revolutions: Retail

1. **Department stores (eighteenth century)**: For hundreds—if not thousands—of years, the retail industry operated on a limited scale that was also highly personal. After making their goods by hand, producers sold them directly to their customers, whom they might know by name. Volume might be low, but customer loyalty helped ensure stability and success.
   In the late 1700s, factories began to open, producing goods *en masse* using steam-powered machines and attracting an influx of workers to cities. Retailers responded to these trends by opening department stores, which offered a variety of products all in one location. In department stores, manufacturers did not come in contact with those using their products; instead, the stores acted as intermediary, or *middleman*, between the maker/manufacturer and the buyer. Without the producer there to guide them, shoppers helped themselves to items, and store assistants helped them find what they needed. The department store model constituted the first disruption, or revolution, in retail.
2. **Shopping malls and plazas (mid-twentieth century)**: The assembly line—the second industrial revolution—enabled the fast production of automobiles. This method of manufacturing made cars and trucks widely available to the general population and decreased their price. Now able to drive to work, many people moved out of cities to the suburbs. This population shift led to a boom in suburban shopping malls and plazas, in which department stores and smaller boutique shops were grouped together in discrete locations—the second retail revolution.
3. **"Big box" stores (late twentieth century)**: Interstate highways ushered in a wave of *big box* stores, or large, no-frills, self-service stores, serving much larger geographic areas. These stores, including Wal-Mart, Aeon, and Auchan, offered a huge selection of deeply discounted goods produced cheaply and efficiently around the world in the automated, globalized third industrial age.
4. **E-commerce**: The fourth revolution—digitization—is occurring in the retail sector in a plethora of ways. Technology now enables consumers to shop for, select, "try on," and purchase goods without entering a store, and receive their items at home, even on the day of purchase.

10.1.2 How Retail Works Today

The ultra-convenience of online shopping has contributed to a decline in visits to *brick-and-mortar* (physical) shops, with some closing their doors for good—a phenomenon known as the *retail apocalypse*. The old paradigms are fading fast.

Technology's impact on retail extends beyond Web sites and virtual shopping carts. Digital is changing not just how people buy, but what they want to happen when they shop, and how they feel. Wearables, mobile computing devices, and Internet of Things appliances and gadgets gratify desires in an instant, and the development of nearly instant delivery of purchased items reinforces that expectation. How can a brick-and-mortar store compete?

Some retailers are struggling to find an answer to this dilemma. While buying a house, owning a car, and other material pursuits have been considered important rites of passage in the past, that paradigm is shifting among younger generations. *Millennials*, people born generally between 1980 and 2000, say they would rather spend money on an experience or event such as a restaurant meal, concert, or trip than on buying a product. Overall, spending on clothing and shoes has declined, while recreation, travel, and dining-out expenditures have increased.

Some retailers, failing to anticipate these changes, remain stuck in old models of competition, continuing to build large stores filled with inventory. As consumer preferences shift from "having" to "doing," however, some so-called big box stores are closing their doors. Increasingly, shop owners are experimenting with ways to offer consumers unique in-store experiences—what futurist Faith Popcorn has called "*consutainment*," a synthesis of convenience, consumption, and entertainment that she sees as the future of retail.

10.1.3 The Fourth Industrial Revolution: Smart Shopping

Technology plays a role in the evolution of shopping, from start to finish.

Virtual assistants such as Google Home or Baidu's DuerOS can tell shoppers which nearby retailers have in stock the product they want—provided that the retailer has placed their store into the database.

Smart beacons, or Bluetooth radio transmitters, can detect when people pass a store, then send notifications to their phones about special offers, perhaps with coupons or discount codes, to entice them inside.

Customers may step through the doors of a shop to find little or no merchandise for sale on site. Instead, they will see items that they can try, order, and pay for in-store, but their purchases will be delivered from a distribution center to their home. This model allows for smaller stores and lower costs to the retailer, and added convenience to shoppers, who no longer need to rifle through racks of merchandise to find the size or styles they need (or to discover that an item is out of stock) or to carry packages as they continue to shop elsewhere.

To entice shoppers to their stores, retailers will create experiences that they can't get online—often using technologies.

- Autonomous vehicles, perhaps owned by a retailer, might pick up shoppers and drive them to a store-sponsored themed event—acrobats in a burlesque show wearing featured apparel or makeup, perhaps.
- WIFI-connected store mirrors or *holograms*, which are three-dimensional images created by light beams, display consumers' clothing choices in 3D, complete with matching accessories, doing away with the need for dressing rooms.
- Robots may move about the floor greeting and helping customers—to find an item in their preferred color or size, to choose the right wine to go with their dinner, or, at an automotive dealership, to select the best car for their needs.
- Virtual reality glasses may allow the shopper to "test-drive" a vehicle or walk through a series of hotels before choosing one to try.

These technologies and others are transforming the shopping experience and enabling retailers to serve customers in new ways. And cloud technology allows them to move smoothly from one stage of the shopping experience to another smoothly, storing their transactions and interactions in one location that they can access anywhere and at their convenience without interruption. This experience is known as *omnichannel*, which means, literally, "all channels."

10.1.4   Benefits of Omnichannel

Omnichannel is the integration of various technology portals to provide a continuous shopping experience for the consumer. This means that a transaction the consumer begins on the phone can continue, uninterrupted, on the Internet, in an app, on social media, with wearables, and during in-store visits, all the way through the purchase. This seamless mode of doing business, although complicated for the retailer, is what customers expect.

In addition to satisfying customer demands, omnichannel retailing also provides value to the vendor in the form of data. Every click, phone call, swipe, and keystroke along the customer journey creates data that may be invaluable for providing personalized, customized services to shoppers. One obvious example is the customization of online ads, tailored according to the viewer's browsing and shopping history. But data holds the potential to transform the shopping experience, as well.

The Mall of America, for instance, uses a chatbot to help visitors plan a personalized itinerary for navigating its 520 stores, 50 restaurants, and entertainment venues.

The big box retailer, Costco, tracks sales of specific items so it can send information about its products to buyers. For instance, when a bacterial infection of peaches, plums, and other stone fruits was discovered, Costco sent alerts to its customers. Such uses of big data can help to generate customer loyalty.

Data might tell an automobile salesperson the types of vehicles a buyer has owned in the past, how much the customer drives, and which features—heated seats?—they use most frequently. It can alert a retailer that a new item in stock holds appeal for certain customers and trigger software to automatically send emails or text messages to those prospects. As these capabilities improve, customers will receive and come to expect, custom-fitted retail experiences tailored just for them.

To accomplish this "customer-centricity," merchants are using *data analytics*, or the analysis of raw data to make conclusions about that information, to help them understand their customers' wants and needs, thus determining how best to fulfill them. As more people incorporate wearables and Internet-connected appliances into their lives, the amount of data generated and the quality of that data will grow. This treasure trove of information, when used correctly, can help retailers put the right offer into their customers' hands at just the right time.

In omnichannel, all channels work together in harmony—using the cloud as a base—to create one continuous experience for the retail shopper. This enhances convenience for the buyer and provides the buyer with more ways to entice and satisfy the consumer.

The typical consumer begins by searching for the product or service they want online. The retailer's Web site will detect the buyer's selections and may have them set aside for their visit. Even while they are in the retail establishment, the consumer will often continue to shop elsewhere using their portable computing devices, looking for bargains or alternative selections.

The retailer, in turn, might engage with these customers' devices by sending push notifications offering instant discounts and deals, up-selling products already in the shopper's cart, or suggesting social media updates for the shopper to post about their experiences.

10.1.5   Going Digital: The Challenges

Especially for established, long-time retailers, these shifts can feel seismic. Making the switch to digital, however, is not optional: To remain relevant to customers, retailers must adapt or die. As the Harvard Business Review puts it: "The answer to disruption is not to double down on a failing model or try to get better and better at things people care about less and less. It's to shift your effort to something they want more." From retailers, people want ultra-convenient, novel, digital experiences.

And yet many businesses have been slow to change. Around the world, many still have no online presence, rendering their businesses all but invisible to online shoppers. Often, they lack personnel with expertise in digital technologies; often, they have no overall digital strategy, one that experts say should include omnichannel, data analytics, and *scaling*, which entails handling rapid growth with a minimum of added resources.

Of course, many other businesses will be trying to do the same—not just locally, but all around the world. With an increase in competition, however, comes a rise in possibilities: For the first time in history, every retailer today has the opportunity to reach prospective buyers everywhere, all over the world.

To take full advantage, some are scaling up, increasing not only the business's reach but also its ability to handle more transactions and to deliver goods and services farther, faster. Scaling up from local to global will entail, among other things, developing an action plan (perhaps part of the digital plan), establishing a team to oversee the process, building relationships with service providers, sales channel partners, suppliers, and distributors, and standardizing operations.

10.2 Case Study: Amazon Alexa

(see Appendix 161–164)

10.2.1 Alibaba

After its founding in 1999, Alibaba grew to become a leading retailer in China and the world, at one point controlling 80% of China's online market.

Not only did it build a formidable online presence with marketplace sites selling to businesses and consumers—and enabling them to sell to one another—but also it offered shopping search engines, cloud computing services, and Alipay, a popular online payment system.

In 2017, however, the retail giant began transforming itself again with an omnichannel model called "New Retail." Blurring the boundaries between online and offline, or so-called bricks-and-mortar, sales, New Retail focuses on providing customized, seamless shopping experiences from phone to tablet to computer to store to home, integrating online and offline shopping (*O2O*).

10.2.1.1 The Technologies

Alibaba's New Retail uses digital technologies to provide an integrated shopping experience tailored specifically for each customer. The company recognized that not everyone wants to shop online—at least, not all the time. For high-ticket items such as cars and kitchen appliances as well as personal items such as cosmetics and hard-to-fit clothing items, many still prefer to "try before you buy." To accomplish this goal, the chain uses:

- **Artificial intelligence**. Alibaba and Bailian Group, a major department store chain, worked to develop new AI-powered technologies and combine their memberships using geolocation, facial recognition, and other systems.
- **Cloud computing**. Alibaba's partnership with fresh-food chain HEMA, which has a curated inventory of 3000 products from 100 countries as well as catered and fine dining offerings, offers an example of how cloud technologies can work to create

an omnichannel experience for shoppers. The partnership allows nearby customers to order online using a mobile app and receive free delivery within half an hour—a feature made possible by logistics partner Canaio. Or they may scan barcodes at the store, pay using the app—linked to payment processor Alipay—and set up delivery at that time.

- **Data analytics**. When customers of Alibaba and its partners shop and buy— in-store or online—digital technologies capture their location, behaviors, and preferences all along the way, allowing Alibaba to personalize their experiences with custom promotions and tailored suggestions.
- **Digital Shopping Applications**. From mirrors that let customers "try on" items virtually to interactive shopping apps, Alibaba's New Retail model emphasizes shopping as an experience as well as a transaction.
- **The Internet of Things**. Sensors using Bluetooth technology track where shoppers are in the store and enact push notifications to their devices alerting to items on sale, offering coupons, making suggestions, and more—adding value for the customer and potentially increasing sales.

10.2.1.2   The Benefits

By moving to a "digital first" mindset and model, Alibaba not only provides customers with the convenience they are increasingly coming to expect from all their shopping experiences but also gains never-before-seen insights into those shoppers.

- **Closing the sale**. By tracking every step of the shopper's journey in the Alibaba space, the business can send reminders to complete the purchase.
- **Customizing the experience**. Knowing a customer's wants and needs before they step foot inside the door—knowledge gleaned from data the shopper created with every keystroke, voice command, and click—enables Alibaba to make special offers tailored to that customer and even make suggestions: a new sauce to try with the fish they've placed into their cart; a handbag to match the dress they're coming in to try.
- **Closing the gaps**. Alibaba's own sizable trove of customer data helps attract partners that can fill gaps in the enterprise's capabilities: developing technologies, financing purchases, and distributing goods around the world.

10.2.1.3   Challenges and Lessons Learned

- **Establishing trust**. When they first began offering goods online, the vendors in Alibaba's marketplace found that Chinese shoppers were reluctant to pay, fearing that the products they ordered wouldn't be delivered. Alibaba established Alipay, an escrow service that held consumers' payments until their orders were delivered.
- **Staying nimble**. The unpredictability of retail as shoppers move from desktops to mobile devices to in-store purchases meant Alibaba had to stay flexible and strive to provide great experiences at every step of the buyer's journey. Its offerings

include online games and contests, chat features that let customers connect directly with vendors to ask questions and get more details on products and services, and in-store events.

- **Holding its technological "edge"**. Keeping pace with consumer demands is not enough for Alibaba to lead in the digital space; it must anticipate those demands, innovate ways to fulfill them, and even dictate needs and desires by developing technologies that are truly disruptive. To help, Alibaba established DAMO Academy, in which leading scientists work on improving the existing technologies and developing new ones.
- **Reaching a fragmented market**. Per-capita use of the Internet is not as high in China as it is in the USA, and so, catering exclusively to online shoppers would limit Alibaba's market and revenues. To diversify and reach offline customers, the organization created LST, a program to help the country's 6.8 million mom-and-pop stores modernize with point-of-sale data collection, payment processing, inventory analysis and suggestions, and an online marketplace on which to sell their wares.
- **Diversifying its offerings**. The more data Alibaba collects, the better it can serve its customers. It collects information on customer preferences and behaviors from partners and stakeholders in a vast ecosystem that includes the grocery, department store, high-end retail, financial, electronics, and logistics and delivery sectors.

10.3   Glossary of Terms

Omnichannel	Digital technologies, hosted in the cloud, that record and store the shopper's interactions and transactions so they can proceed smoothly and easily from start to finish, whether on the phone, the Internet, in an app, on social media, with wearables, and during in-store visits, all the way through the purchase.
Middleman	The intermediary between the manufacturer and the consumer.
Retail apocalypse	Closing of physical retail stores predicted in response to the growing popularity of online shopping.
E-commerce	Online shopping.
Consutainment	A synthesis of convenience, consumption, and entertainment, seen by some as the future of retail.
Holograms	Three-dimensional images created by light beams.
Scaling	A business's ability to handle rapid growth with a minimum of added resources.

10.4   Questions

1. How are digital technologies changing the buying and selling experiences for consumers and retailers?
2. What are the four industrial revolutions in retail, and how has each transformed the sector?
3. What are three technologies retailers are using to make in-store shopping more enjoyable? Can you think of some others that they should try?
4. What are two benefits to consumers that omnichannel technologies offer? What are two benefits to retailers?
5. Describe Alibaba's "New Retail" initiative, and which technologies are important to its success.
6. What are some of the benefits that Alibaba is gaining from omnichannel?
7. What is the role of data in Alibaba's business model? Why is this important?

Chapter 11
Transportation, Travel, and Tourism Evolution

11.1 Transportation, Travel, Tourism, and Big Data

The transportation, travel, and tourism sector—abbreviated in this chapter to 3T—is making a subtle but dramatic shift. The industry led the way into the digital age with *e-tourism*, the use of digital media to assist with every transaction in the 3T value chain. E-tourism allows travelers to seek, find, and procure experiences with a few keystrokes or clicks. Now, the industry is undergoing another transformation with *smart tourism*, offering not only an array of options via digital media but also making personalized suggestions and even guiding the consumer's choices.

Imagine this scenario: You are a college student planning to visit a city in Europe. You enter the city's name into your mobile travel planner, which suggests lodging within your budget near an area known for a lively nightlife scene, as well as popular and inexpensive places to eat. It scours the Internet to find you the best prices on airline tickets and suggests Metro routes to your lodging along with a public transit pass good for the duration of your stay, which you can purchase instantly. It also informs you of a hat shop in the area and a museum featuring works by one of your favorite artists.

© The Editor(s) (if applicable) and The Author(s), under exclusive license to Springer Nature Singapore Pte Ltd. 2020
J. R. Reagan and M. Singh, *Management 4.0*, Blockchain Technologies, https://doi.org/10.1007/978-981-15-6751-3_11

How does your device know that you are a student on a budget who enjoys good food and nightlife? How does it know that you wear hats and appreciate a certain artist? It knows because of your past digital interactions and transactions, each of which generates digital information, or *data*.

Data is the engine that drives the fourth industrial revolution. None of the technologies explored in this textbook would work without data. And among the industries discussed, smart tourism, with its innate focus on customer experiences, will perhaps be the most dependent of all on data and *data analytics*, which is the process of drawing insights from raw information to add value.

In other words, tourists are not the only ones benefitting from the connected age. Nor, however, are they passive consumers, but are now contributors to the smart tourism ecosystem. Every choice travelers make becomes data to be sifted, analyzed, and used to inform decisions about services and products provided to them. Every customer review or rating becomes an opportunity to refine and improve—all in the name of enhancing competitiveness and decreasing uncertainty. And all the data generated informs algorithms used to personalize suggestions and offers for tailor-made experiences and, hopefully, improved customer loyalty.

The United Nations World Trade Organization defines *tourism* as "a social, cultural and economic phenomenon which entails the movement of people to countries or places outside their usual environment for personal or business/professional purposes."

Smart tourism uses data and technology to inform, guide, and enhance the tourist's choices. Its components include *smart destinations*, digitally connected locales such as cities and cultural/historic attractions; *smart businesses* in the transportation, hospitality, and food and beverage industries; and *smart tourists*, travelers willing and able to use and perhaps even contribute to that ecosystem.

Smart tourism may use some or all of the following technologies to both augment and glean from the tourist's experience:

- Beacons,
- Sensors,
- The Internet of Things,
- Bluetooth,
- RFID,
- GPS,
- Augmented reality (AR),
- Internet/Web sites,
- Apps,
- Mobile devices (smartphones, tablets, watches, etc.),
- Artificial intelligence (AI),
- *Collective intelligence*, in which technologies share information and learn from one another.

To deliver a truly smart experience—one informed by, and participating in, the connected world—destinations must be linked and coordinated, sharing information not only with the consumer but also with one another, all with the customer in mind.

11.1.1   The Four Industrial Revolutions: Transportation, Travel, and Tourism

- **Steamships (nineteenth century)**: The early to mid-nineteenth century saw the invention of steamships and steam engines for train travel, enabling for the first time large-scale, efficient cross-continental and inter-continental travel.
- **Automobiles (twentieth century)**: The combustion engine, electrification, and assembly-line production introduced around the turn of the twentieth century made automobiles widely available and affordable, increasing personal mobility as never before. The invention of the airplane and rapid advances in aviation led to safe, affordable air travel.
- **The Internet (1990s)**: Advances in digital technologies and electronics, especially personal mobile computing devices and the Internet, paved the way for online reservations and information about tourist destinations, e-tickets, and customer-generated reviews, essential components of e-tourism.
- **Smart tourism**: Advances in data analytics and artificial intelligence enable smart tourism, increasing convenience for the tourist/traveler, and certainty and competitiveness for 3T businesses and destinations.

11.1.2   Tourism and Travel: Yesterday and Today

Before the Internet, making travel plans was a time-consuming and often difficult and onerous task.

A tourist usually either hired a travel agent to make airline and hotel reservations or did so themselves on the phone after seeking information from written travel guides available in bookstores and libraries.

Airline tickets were issued in paper form, usually sent through the mail, and losing them meant paying a substantial fee to get them replaced.

At their destination, travelers relied on the information and maps provided by hotel concierges or the guidebooks they carried with them to get information on restaurants and cultural sites, including street addresses, hours of operation, attractions, and admission fees. Updated annually, these guides quickly became obsolete and had to be replaced for use on subsequent trips.

The Internet ushered in the age of e-tourism, with transportation, lodging, and destination information and bookings all available online. E-tickets replaced paper ones, relieving anxiety over forgetting or misplacing tickets. Booking sites offer discounts, and travel Web sites provide information and customer-sourced reviews of amenities including hotels, restaurants, and cultural/historical sites as well as photos, maps, and travel tips.

11.1.3 Industry 4.0: Smart Transportation, Travel, and Tourism

The convergence of "smart" technologies and digital innovation is ushering in a new era for 3T. No longer do tourists rely solely on Web sites to get information about their destination. Today, travelers' devices, working in conjunction with sensors, beacons, and the Internet of Things, bridge the physical and digital realms to not only help with planning and reservations but to enhance the travel experience during the trip.

For instance, instead of getting inspiration from a Web site or social media to visit a place, tomorrow's traveler may view a destination using virtual reality. They may request information and help from their virtual personal assistant, which finds options and even books the trip. If the traveler switches to the Internet, they can complete the transaction on-screen without interruption, perhaps with the help of a chatbot.

At the airport or train station, there are no long lines or check-in kiosks, but travelers are automatically checked in when they walk into the facility. If they miss their flight, they get rebooked automatically. Augmented reality maps and/or robots guide them to where they need to go, and RFID tracking tells them where their luggage is at all times.

At the destination, an autonomous vehicle picks up the traveler curbside and takes them to their hotel. They are automatically checked in when they enter the hotel, and their mobile device becomes their room key. AR maps lead them to their room, and robots take care of their luggage. If it's mealtime, their device offers recommendations on nearby places to eat and reserves a table upon request.

To facilitate all this smart interaction, businesses and destinations are providing free apps, WIFI, *near-field communication* (NFC), a form of communication that uses very-short-range radio transmission to exchange data among devices in close proximity to one another (such as when we use our phones to make a credit card payment); augmented reality displays, virtual assistants, and even smartphones to help travelers find and enjoy experiences. Doing so helps them not only attract and gather positive ratings from consumers but also enable them to collect data from users' experiences and interactions, as well.

11.1.4 Benefits of Smart Tourism

Data provides a plethora of benefits to the 3T industry, potentially enabling purveyors to:

- Plan for "boom" and "bust" seasons,
- Anticipate and resolve problems quickly,
- Operate more efficiently,
- Promote their services,
- Innovate,
- Attract investments,
- Improve communications,
- Differentiate themselves from the competition,
- Control crowds,
- Increase sustainability,
- Improve customer service.

11.1.5 Smart Tourism: The Challenges

The transformation from e-tourism to smart tourism will require a shift in thinking among those in the travel and hospitality industry, away from competition to collaboration. That's because technologies are only as smart as the information they have. For the high-quality user experiences, hotels and hostels, destinations, transportation services, booking services, restaurants and bars, and other entities will need to work together, sharing their data and moving from a "value chain" model to "ecosystems" in which each is dependent upon, and nurtures, the rest.

Some investment in technologies will be necessary, as well, requiring a digital strategy for organizations wishing to participate. Far beyond throwing up a Web site and signing on to a booking service, destinations and businesses will need to use sensors, NFC, the Internet of Things, robots, Bluetooth, AR, AI, and/or other technologies to satisfy consumers' expectations. Since these technologies all provide valuable data as well as help to attract and retain customers, they may be seen as investments.

Processing, storing, and analyzing data may be the biggest challenge, however. *Open data*, or that which is collected, anonymized, and made publicly available, as well as *big data*, or the voluminous raw data businesses collect, must be processed, stored, analyzed, and organized to be of any use. Developing systems and finding people with the expertise and ability to make intelligent use of the data generated is key—and a problem that technology will probably resolve in time.

In the race to "get smart," though, the transportation, travel, and tourism industries may find it best not to eradicate the human touch from the equation. Studies show that a friendly face can be much appreciated in a strange land, and travel is, first and foremost, a service business. Front-facing employees will still be needed to welcome weary travelers and make them feel at home and to answer their questions and help solve their problems. They will need to understand and be able to interact with the many technologies customers are experiencing, as well.

Smart tourism is already beginning to transform the transportation, travel, and tourism industry. For the traveler, it will provide customized, optimized options, decrease the uncertainty and confusion inherent in the travel experience, and provide reassurance and convenience—increasing the enjoyment that tourists expect. For destinations—businesses as well as public entities—it can reduce or even eliminate unpleasant surprises and provide the information needed to improve the delivery of services, fostering loyalty among the existing customers and attracting new ones.

11.1.6   The Benefits

Since the beginning of the Smart Dubai initiative, the city has seen its visitor numbers grow year by year, bringing it ever closer to its stated goal of being number one in tourism.

And the project's "Happiness Meter" app enables the government to know whether it is fulfilling its top priority—its "Happiness Agenda." "The Happiness Meter will enabling users at many public and private sites throughout the city to click on one of three faces—frowning, neutral, and sad—the meter interactively measures customer satisfaction with digital experiences—and tells the government how close it's coming to its ultimate goal of 'making Dubai the happiest city on Earth.'".

11.1.7 Challenges and Lessons Learned

To establish and maintain the plethora of technologies enabling its smart tourism initiative, the Dubai government worked with a number of private partners including the TenCent International Business Group, the Chinese technology conglomerate behind WeChat and Weixin (more than 1 billion active monthly users altogether).

Working in cooperation with business and technology providers enabled the city to enhance Chinese visitors' digital experiences by using artificial intelligence, the Internet of Things, cloud computing, and big data, and to collect and share tourist-generated data for Dubai's analysis.

Staying not just current with technologies but ahead of them, anticipating users' needs and desires, can be especially challenging for a city that entertains so many millions of visitors from around the world. Innovation is key, and to help, Dubai's Department of Tourism and Commerce Marketing launched the Future of Tourism Challenge, in which technology startups have competed for funding and partnerships. The 2018 contest focused on artificial intelligence, digital ecosystem (measuring the satisfaction of users and personalizing their suggestions and experiences), VR/AR, and blockchain.

11.2 Case Study—Dubai, UAE

(See Appendix 184–186)

11.3 Glossary of Terms

Tourism	A social, cultural and economic phenomenon which entails the movement of people to countries or places outside their usual environment for personal or business/professional purposes.
E-tourism	The use of digital media to assist with transportation, travel, and tourism transactions.
Smart tourism	Tourism that uses data and other technologies to suggest personalized experiences and help guide the consumer's tourism choices.
Data	Digital information.
Data analytics	The process of drawing insights from raw information to add value to a service, product, or experience.
Smart destinations	Digitally connected locales such as cities and cultural/historical attractions.

Smart tourism businesses	Businesses in the transportation, hospitality, and food and beverage industries that use smart tourism.
Smart tourists	Travelers who participate in smart tourism who may contribute to it with reviews and social sharing.
Near-field communication (NFC)	A form of communication that uses very-short-range radio transmission to exchange data among devices in close proximity to one another.
Collective intelligence	Technologies' sharing of information, and learning from one another.
Open data	Data that is collected, anonymized, and made publicly available.
Big data	The voluminous raw data businesses collect.
Expatriate	Someone who lives outside their native country.

11.4 Questions

1. What's the difference between e-tourism and smart tourism?
2. Write your own "smart tourism" scenario, in which you plan a trip and technologies assist you every step of the way.
3. What is the role of data in smart tourism?
4. What are some benefits to the transportation, travel, and tourism industry of collecting and analyzing information about tourists and their experiences?
5. What are some uses that you can think of for collective intelligence, in which technologies share information and learn from one another?
6. Describe the three industrial revolutions before smart tourism, and explain how smart tourism is revolutionizing the transportation, travel, and tourism industry.
7. What are the four key technological components of Dubai's Smart Dubai project?
8. Name one challenge Dubai faces in meeting its goals—using smart digital technologies to increase the happiness of those visiting and living in the city and increasing tourism—and describe how the city is meeting that challenge.

Chapter 12
Social Revolution

12.1 Social Impacts of Industry 4.0

Never in human history has humanity experienced such rapid, constant, and pervasive change as is occurring at this very moment. The fourth industrial revolution is disrupting every facet of business and personal life at a dizzying pace, all around the world. Manufacturing, politics, health care, education, media, work, home—even dating and relationships—all are experiencing dramatic upheavals in norms and expectations, in potential and possibilities, as humans become more intricately connected with machines, and machines with one another.

Are these changes positive for society, or negative? Intense debate has accompanied the transformation our world is experiencing and continues today.

Optimists point to the advances in medicine promising longer, healthier lives, for instance, and conveniences only imagined, before now: cars that drive *us*; robots in our factories and homes; appliances that do our bidding at a single word, or a wave of the hand.

Critical science, on the other hand, focuses on technology's adverse effects: losses of jobs, privacy, and security, to name a few. Among the world's most esteemed thinkers, there seems to be little agreement on whether artificial intelligence, the Internet of Things, and other connected technologies bode well or ill for humankind.

12.1.1 The Benefits

Digital technologies are bringing about many changes for social good:

- Medicine has made dramatic progress in recent years, including genetic research, *precision medicine*, or the tailoring of medical treatments to individuals based on their unique physiologies, health monitoring via wearables, the use of virtual reality to aid in psychological therapies, and more.

- The emergence of the online marketplace in the retail sector has opened up opportunities worldwide for people to start their own businesses, and for consumers to find and purchase their goods and services.
- Connected campuses and online courses have expanded access to education, much of it free.
- Connected services and infrastructures have resulted in "smart," sustainable cities and societies, reducing fossil fuel use and potentially helping to reduce the greenhouse gas emissions causing climate change.
- On-demand access to information and social networks for sharing it has exploded the potential for acquiring knowledge and empathy for others.
- Connected modes of transportation, communications, consumer goods, and other aspects of everyday life save users time and money.

12.1.2   The Negative Effects

Technology has unintended consequences, as well—and the rapid pace of innovation and change has made it often difficult to address issues proactively, or even to foresee them. With so many technical, industrial, and social innovations wreaking continual change, the ability of individuals and institutions to adapt and overcome the resulting threats to human identity, social stability, and economic security is being tested. Concerns include:

- **Loss of privacy**. Smart technologies allow us to access and share information as never before, but it also allows other people and institutions to access and share information about us. Technologies such as *electronic profiling*, building personal profiles of individuals using information from a variety of electronic sources— credit card transactions, property records, online purchases and browsing history, social media posts, public records, and more; *facial recognition*, the automated matching of a person's photo with their identity, often without the individual's knowledge or consent; and *Uberveillance*, or electronic tracking and information gathering combined with personal data to continuously track a person's location. Many fear that this widespread gathering, surveillance, and disclosure of personal information stand to eradicate the very existence of privacy.

- **Loss of employment**. Since at least the first industrial revolution, technologies have displaced humans in the workforce, performing tasks more quickly and at a lower cost than people can do. Connected technologies, especially artificial intelligence, have accelerated this trend, replacing people in factories and in transportation, shipping, warehousing, agriculture, and many other professions. Less-skilled workers are the most vulnerable, but some predict that as AI becomes more responsive and complex, machines will displace many higher-level workers, as well. As the job market shifts, implications are enormous and could result in large-scale unemployment unless people can be retrained to work in industries such as technology and cybersecurity, where demand will continue to grow.

- **Loss of safety**. Even the most revered thinkers continue to debate whether AI might pose a threat to humanity itself. Already computers can perform calculations and other tasks that humans cannot: Is it possible to invent machines that can outsmart us, as well? Some scoff at the notion, but others have issued dire warnings. The late English physicist Stephen Hawking predicted, "Unless we learn how to prepare for and avoid the potential risks, AI could be the worst event in the history of our civilization. It brings dangers, like powerful autonomous weapons, or new ways for the few to oppress the many." Facebook founder Mark Zuckerberg has said that scientists don't understand how some types of AI learn and make decisions.

- **Security**. *Cybercrime*, or criminal activities carried out via computers or the Internet, stands to increase with every connected device. Each home appliance in the Internet of things, each connected means of transportation, each robot and drone and connected wearable device, offers another portal through which hackers could potentially access our data, bank accounts, workplaces, homes, and even our bodies, disrupting business, stealing assets, and even endangering national security.

- **Sustainability**. Every new computing device needs power to work, be it wearables, sensors, robots, drones, or something else. An added demand for power raises the demand on our planet's natural resources while also potentially increasing the release of greenhouse gases known to cause climate change. In addition, these devices are made from valuable resources and highly engineered materials, including precious metals, plastics, and glass, all of which use energy in the extraction and manufacturing process. Transporting devices from factories to consumers use energy, as well.

12.1.3 Smart Societies

As the fourth industrial revolution disrupts virtually every sector of life, so will industries, governments, and societies need to adapt to and even help direct the cultural shifts that will ensue, to create *smart societies*, or societies using digital technologies for citizen well-being, strong economies, and more effective institutions including government and education.

One way to achieve these goals on a global level is through collaboration and sharing of information. Governments, businesses, and citizens can learn from one another what works and what doesn't, adopting and adapting solutions for their own societies' good.

Recommendations for how to achieve the paradigm shifts needed to achieve truly smart societies include:

- Educators could ensure that an education in *STEM*—science, technology, engineering, and math—includes humanities courses such as ethics. Today's technology students are tomorrow's innovators, and their values will help shape not only our digital future but also our societal well-being.
- Policymakers and technologists might work together to understand the potential negative effects of digital technologies on society and enact policies, practices, and laws to counter those effects.
- Businesses could institute social responsibility and sustainability initiatives, drawing on their institutional knowledge and experience to direct the use of technology in ways beneficial to society.
- Businesses and governments could work together to investigate and counteract cybercrime.
- All stakeholders, including governments, industries, schools, non-profits, media, and individuals, could gather for roundtable discussions to agree on the most desirable outcomes of smart societies and how to achieve them.

Technology is neither inherently positive nor negative, but merely a tool—a very potent one. How humans use this tool will determine the nature of its impact on society. People create technologies, and people will determine whether they advance human progress or hinder it. In the connected age, will we use our powers for good or evil?

12.2 Case Study: Society 5.0 Japan

(see Appendix 169–173)

12.3 Glossary of Terms

Critical science	A model of science in which **scientific** methods are used to critique the adverse consequences of technological development.
Precision medicine	The tailoring of medical treatments to individuals based on their unique physiologies.
Electronic profiling	Building personal profiles of individuals using information from a variety of electronic sources—credit card transactions, property records, online purchases and browsing history, social media posts, public records, and more.
Facial recognition	The automated matching of a person's photo with their identity, often without the individual's knowledge or consent.
Uberveillance	Electronic tracking and information gathering, combined with personal data to continuously track a person's location.
Cybercrime	Criminal activities carried out via computers or the Internet.
Smart societies	Societies using digital technologies for citizen well-being, strong economies, and more effective institutions, including government and education.
STEM	Science, technology, engineering, and math.

12.4 Questions

1. What are some of the viewpoints regarding the social impacts of the connected age? Do you agree with any of them? Why or why not?
2. Name three supposed benefits of connected digital technologies, explaining why they are thought to be good and whether you agree, providing reasons for your opinions.
3. Name three effects of digital technologies that are seen as negative. Do you agree that these effects are bad for society? Why or why not?

4. What do you think of Stephen Hawking's prediction regarding artificial intelligence? Why?

5. What are some ideas for achieving "smart societies"? Can you add one or two of your own?

Chapter 13
Future Revolution

13.1 What's Next: Revolutionary Changes

In addition to transforming daily life, the fourth industrial revolution is disrupting business down to its most fundamental structures, requiring entirely new models of operation. While the first, second, and third revolutions also wrought significant changes, none matches Industry 4.0's speed and scale. The "connected age" is completely transforming how business functions in terms of customer engagement, production, research and development, delivery, sales, marketing, service, and more.

By interconnecting people, machines, buildings, and devices into a vast, ever-expanding network, Industry 4.0 is creating new challenges and opportunities in every sector.

As they seek to compete, enterprises face imperative revisions in their:

- *Business model*, defined as a design for the successful operation of a business, identifying revenue sources, customer base, products, and details of financing,
- *Operating model*, which is how a business uses people, processes, and technology to fulfill its business strategy.

Some of the business requirements of this new age are:

- A "digital first" mindset,
- A commitment to creativity and continually innovating,
- A shift in focus to "customer-centric,"
- An inclusive, collaborative approach that values suppliers and vendors not merely as links in the *value chain*—defined as all the activities from concept to delivery that add value to a product or service—but as partners in a mutually dependent business *ecosystem*, a cooperating network of suppliers, distributors, customers, vendors, and even competitors.

J. R. Reagan and M. Singh, *Management 4.0*, Blockchain Technologies, https://doi.org/10.1007/978-981-15-6751-3_13

13.1.1 A Different Revolution

Every industrial revolution in history has changed the way businesses operated and, by extension, the way people lived their lives. Unlike Industry 4.0, however, the first three revolutions happened gradually and required strategic shifts in business practices rather than complete overhauls.

- **The first revolution**, powered by steam and machines such as the cotton gin, turned the world's agrarian societies into ones dominated by large urban centers. It moved the production of goods, beginning with textiles, from the hands of skilled craftspeople to factories. In doing so, this revolution created new consumer markets for these goods, making them available to more people at lower prices than before. At the same time, the development of the steam engine revolutionized travel and shipping, moving people and products across continents and over oceans in a matter of weeks rather than months, and enabling international exports and imports of goods, services, and culture.
- **The second revolution** built on the first. Electricity replaced steam power, enabling automation in the manufacturing process, which facilitated mass production. More goods entered the market using less labor and at a lower cost than ever before, and an increase in factory jobs provided people with the income to purchase manufactured items—including the automobile, which had previously been handmade and prohibitively expensive for most. The development of the combustion engine enabled vehicles to go farther, faster, lending unprecedented mobility to people and products and expanding markets even farther.
- Globalization, the next logical step, occurred with the **third industrial revolution**, stimulated by the invention of the Internet and the availability of computers. For the first time, businesses could offer their products and services to a worldwide market and employ workers in remote locations, as well.

Each of these transformations occurred over time, but at an increasingly accelerated pace. The first industrial revolution, for instance, began around 1760; the second, about 110 years later, around 1870; the third, less than 100 years later, in the mid-1900s, and now, in the early twenty-first century, a fourth industrial revolution, the "connected age," is already occurring.

In the **fourth industrial revolution**, mobile devices, including wearables and sensors, the Internet of Things, and the cloud, are joining our bodies with our possessions and places, including our homes and offices. At the same time, these technologies are also linking places and things with one another.

The data these devices generate enables artificial intelligence, which analyzes and synthesizes that data to make autonomous decisions, supporting such technologies as the smart grid, smart cities, and autonomous vehicles. It also gives businesses the ability to tailor their offerings to individual customers, while the always-on nature of these connections provides consumers with the convenience of being able to shop for anything they desire from wherever they happen to be, whenever they wish to buy.

13.1.2 Challenges

The ubiquity of digital technologies and shifts in customer expectations require changes in the following areas:

- **Product Design.** As more items connect to the cloud, designs will be more complex, adding digital elements to physical ones. The data these products collect will enable continual upgrades (*evergreen* design) as opposed to periodic ones, as well as frequent and rapid redesigns and improvements. Computer systems engineering, remote service and *predictive maintenance*, in which devices determine in advance when they will need servicing, security, and personalization/customization will all need to be incorporated into product design. Crowdsourcing, data analysis, and feedback enable the customer to become a part of the design process.
- **Sales and marketing.** In the connected age, products are not inert things but active sensors, collecting data that measures its value to consumers and enabling sales and marketing teams to better divide their customers into segments for more effective, targeted marketing. Sales efforts will shift away from one-time product placement to upgrades, replacement products, and services offered for an extended period over the product's lifespan.
- **Manufacturing.** Physical and digital processes will converge as smart, connected equipment, and robots work together for factories automated from "top floor to shop floor." Physical components and assembly will become simpler as more products rely on software to operate, and design changes will become possible later in the process, even post-manufacturing, while the product is in use.
- **Service.** Instead of periodic maintenance (whether a product needs it or not), and repairs after something has broken, service delivery will happen in a predictive, proactive manner and will often occur remotely. Customers will be able to analyze how they use their products and make changes to prevent problems and extend their useful life, and service providers can do the same to increase service efficiency and better manage warranties.
- **Human resources.** Although digital advances threaten to displace lower-skilled workers with robotics and other forms of artificial intelligence, many skills that only humans can do are increasingly in demand. Businesses will need to recruit and retain workers with IT skills and even train the existing employees to move into these jobs.
- **Competition.** Price will no longer be the primary consideration for consumers wishing to buy a product; instead, the quality of the customer's experience, which incorporates user-friendly design, customization, and services, will be paramount.

Consumers have come to expect personalized digital experiences on demand, and businesses are scrambling to not only keep up but also anticipate their customers' wants and needs. Doing so requires *agility*, defined as the ability to quickly and nimbly change business practices and product offerings—important for success in a world where technology advances at a rapid and continual pace. Agility also means these businesses must emphasize and enable innovation and change, as customers increasingly value experiences over material things and shift loyalties to brands that set trends rather than follow them.

Frequently introducing new products and updates represents a major shift for businesses in most sectors. The auto industry, for example, has traditionally rolled out new models of cars and trucks on a yearly basis. With vehicles increasingly connected to the digital network, however, annual model updates are no longer enough: When an improvement is available in the digital technology used in the machine, automakers must update it in the cars and trucks already sold as well as those not yet sold.

Simply pushing out patches and updates is far from enough to satisfy today's demanding customer, though. New paradigms require industry players to keep a close eye on consumer demands and introduce upgrades and new features that solve their problems regularly and frequently—or someone else will. Becoming agile requires, for most businesses, complete overhauls in structure and operations.

The roads to agility are many. Essentially, agile business practices mean smaller teams of people working to innovate in key areas, adapting to customers' ever-shifting priorities, bringing new products and features to market more quickly, and reducing risk. Components may include:

- *Scrum*, in which teams of people from multiple areas of the business work together, using creative and adaptive thinking to solve problems,
- *Lean development*, continually working to eliminate waste in favor of "lean" processes, and
- *Kanban*, which concentrates on reducing the amount of time it takes to develop and introduce a new product or update, minimizing time spent on micromanaging and freeing senior managers to focus on higher-value tasks.

These models can help remove obstacles to progress, but in many cases, will require fundamental shifts in how organizations function.

Other challenges that businesses face in the connected age include:

- Weathering more volatile markets,
- Innovating for shorter product lifecycles,
- Servicing more complex products,
- Navigating global supply and distribution networks.

Many of the problems that technologies create, however, can also be resolved or at least mitigated by those same technologies. For instance, technologies such as blockchain may increase security. Virtual and augmented reality can help businesses train workers to perform tasks that AI cannot.

13.1.3   Opportunities

The combination of connectedness via the Internet of Things and the cloud, smartness via artificial intelligence, customer insights via data analytics, and other advantages offered by digital technologies can help businesses improve their practices and adapt to the new era. Examples include:

- **Smart systems**. To respond more quickly to fluctuations in supply and demand, retailers and manufacturers are using artificial intelligence to enable their systems to communicate and make decisions in the moment. For example, when a factory faces shortages of a needed component, AI-driven systems can contact suppliers to find the part and place the needed order—or, conversely, halt or slow orders where demand is low. At the same time, autonomous systems can assign workers where they are needed at any given time. This melding of "top floor" and "shop floor" can not only increase efficiency and prevent shortages but also boost productivity and lower costs.
- **Data-driven decisions**. Small, diverse teams focused on innovating new and better goods and services have been shown to help businesses reduce the time it takes to bring new products to market and gain an advantage over competitors. The data devices collect about consumer uses and habits can help these teams better meet and anticipate their customers' needs.
- **Customer engagement**. Social media, consumer ratings, crowdsourcing, and other forms of digital communication can help businesses better engage with their customers and make them feel valued in today's "customer-centric" environment. Soliciting customers' ideas can also help enterprises innovate changes to customize and personalize their products and services, improve the products and services they have, and create new ones that will delight their customers and attract more business.
- **Sustainability**. Digital technologies use energy, taking a toll on an already-taxed environment. Data collection and analytics help businesses add value for customers. Also, digital solutions such as connected solar and wind power can help companies to manage their energy use better, perhaps resulting in reductions and cost savings.

13.1.4 Industry 5.0: The Return of the Human

Barely into the fourth industrial revolution, some are already anticipating *Industry 5.0*. Many predict that the fifth industrial revolution will see robots and other forms of artificial intelligence co-existing and working in harmony with humans.

This new era may see a true convergence of humans and machines—literally and figuratively. Instead of our using smartphones loaded with applications, technologies will live on our bodies, with virtual assistants murmuring directions in our ear, suggesting restaurants for dinner, making purchases on our behalf, and much more.

But the most paradigm-shattering changes will occur in the workplace. Industry 5.0 is expected to see the transformation of the fourth revolution's "cyber-physical" manufacturing plants—those using digital technologies to operate factories with minimal human involvement—into "human-cyber-physical" systems.

In this new world, data gets collected and processed at faster and faster speeds. Machines and robots use the information to make decisions based on programmed algorithms and their own databanks containing "memories" of their past actions and outcomes.

Far from marginalized as some predict, humans hold center stage in this new revolution. Machines serve us, not the other way around.

In this new paradigm, people will work alongside collaborative robots, or "cobots," teaching them to do their jobs and correcting them when they make mistakes. Machines will perform the most repetitive, menial, and dangerous tasks, while people will use our intricate, flexible brains to make high-level decisions. For example, humans will design products and processes, perhaps using a "digital twin," or a virtual copy of the factory where the product gets made or the environment where it will be used.

Along the way, the factory's ability to communicate directly with customers will enable it to customize and personalize every product according to individual needs and desires.

Some of this is already occurring. Much of it has yet to transpire. As we peer into the near future and try to imagine the new world that will so quickly emerge, ask yourself: How can I be ready? Or even: How can I lead the way?

13.2 Glossary of Terms

Business model	A design for the successful operation of a business, identifying revenue sources, customer base, products, and details of financing.
Operating model	How a business uses people, processes, and technology to fulfill its business strategy.
Value chain	All the activities from concept to delivery that add value to a product or service.
Business ecosystem	A cooperating network of suppliers, distributors, customers, vendors, and even competitors.
Evergreen design	Continual upgrades in product and service design, enabled by the collection and analysis of user data.
Predictive maintenance	Items' ability to determine in advance when they will need servicing.
Agility	The ability to quickly and nimbly change business practices and product offerings.
Scrum	An agile business practice in which teams of people from multiple areas of the enterprise work together, using creative and adaptive thinking to solve problems.
Lean development	Continually working to eliminate waste in favor of "lean" processes.

Kanban   Reducing the amount of time it takes to develop and intro-
                           duce a new product or update, minimizing time spent on
                           micromanaging, and freeing senior managers to focus on
                           higher-value tasks.
Industry 5.0   The next industrial revolution, seen as an era in which
                           robots and other forms of artificial intelligence will co-
                           exist and work in harmony with humans.

13.3 Questions

1. What are some of the actions that businesses will need to take to adapt to the
   connected age successfully?
2. Briefly describe how each of the previous three industrial revolutions transformed
   business and markets. How might Industry 4.0 do the same?
3. Name three areas of business that will need to change in the connected age,
   describing some of these changes.
4. Describe business agility, listing some of its aspects, and explaining why it is
   important.
5. What are some ways that digital connectedness can help enterprises? Which do
   you see as the most critical, and why?

Case Study—Industry 4.0

Germany's Industrie 4.0

The government of Germany wants to lead the way into the fourth industrial revolution. Intending to establish the nation as a world leader in manufacturing, government ministries have established Industrie 4.0, a strategic initiative funding technological research, forming industry networks, and standardizing technologies.

The Technologies

Funded by the Ministry of Education and Research and the Ministry for Economic Affairs and Energy, Industrie 4.0 envisions improvements in manufacturing processes using such technologies as:

- **Data**: Collecting and analyzing data can help improve quality control by identifying weaknesses and flaws in the manufacturing system and finding solutions.
- **Autonomous vehicles**: Logistics vehicles operate themselves, moving goods automatically and intelligently from one production point to the next.
- **Cyber-physical systems**: In tomorrow's fully automated "smart" factories, machines oversee production for increased productivity, less downtime, and lower personnel costs.
- **Robotics**: Equipped with sensors, cameras, and digital connections, automated, autonomous robots can work on the factory floor, performing many tasks previously accomplished by humans without human error or fatigue.
- **Production line simulation**: *Digital twins* of production lines before they are established allows engineers to safely test them for efficiency and effectiveness before they are installed.

© The Editor(s) (if applicable) and The Author(s), under exclusive license
to Springer Nature Singapore Pte Ltd. 2020
J. R. Reagan and M. Singh, *Management 4.0*, Blockchain Technologies,
https://doi.org/10.1007/978-981-15-6751-3

The Benefits

The government is not the only investor in Industrie 4.0. A number of research centers, a consortium of industry stakeholders, and private industries are collaborating on the initiative.

The potential for gain is high: One study estimated that Industrie 4.0 could generate 79 billion euros' growth in the country by 2025 in six sectors: chemical engineering, automotive, mechanical engineering, IT and communication, electrical engineering, and agriculture.

- **More effective and efficient workplaces**. Factories can operate more smoothly and at lower cost using connected technologies, as can businesses in many research and developments in one sector benefits all.
- **Enhanced reputation as a world industrial center**. Manufacturing already makes up a large portion of the German economy, making it one of the most competitive nations in the world for industry. By looking to the future, the country hopes to maintain that status.
- **A competitive technological edge**. Industrie 4.0 is a private–public partnership, with research and development funded by the government as well as business dollars. Technologies developed and enhanced under the initiative may ultimately benefit these businesses as well as the German economy.

- **A boosted national economy**. Increased productivity, improved customer services, a workforce trained to perform in the digital age, and standard-setting technology products may stimulate investments in German business, industry, and infrastructure, creating jobs and revenues and enhancing quality of life. The added gross value nationwide generated by Industrie 4.0 is expected to average 1.7% per year.

Challenges and Lessons Learned

Security. Securing data is difficult enough, and the stakes are high for privacy and proprietary information. Securing all the devices used in a smart factory presents a unique set of challenges, with failure potentially causing breakdowns and even danger to human safety.

Infrastructure. Particularly in rural areas, Germany suffers from slow and even spotty Internet service. Without fast broadband, engineers and manufacturers may not even realize the possibilities the digital age offers to manufacturing.

Standardization. Although this is a goal of Industrie 4.0, many technology products are still available using many different platforms and interfaces. Smaller companies and suppliers, in particular, will have a harder time converting to digital if they have to continually buy new technologies and train workers in how to use each new software product.

Case Study—Agriculture

The Hands-Free Hectare

Researchers in England used drone technology embedded in agriculture machinery to drill, sow, tend, and harvest without ever stepping foot on the soil. Hands-Free Hectare in 2017 produced the world's first crops grown completely with autonomous technologies, using drones, sensors, computers, and a modified tractor and combine.

In the near future, the researchers say, all farms will operate this way.

The project, at Harper Adams University in Shropshire, used lightweight equipment with an eye toward alleviating soil compaction, which can cause water to run off soil instead of soaking in, and hinder root growth. With £200,000 from the British government and private partner Precision Decisions, they added a sprayer boom, a seed drill, and drone technology—cameras, sensors, lasers, GPS, and more—to a small tractor and a 25-year-old combine harvester. Rather than buying expensive autonomous machines, the teams opted to modify existing equipment, in part because of their small budget, but also to show that farmers can use the techniques without breaking their bank accounts.

© The Editor(s) (if applicable) and The Author(s), under exclusive license to Springer Nature Singapore Pte Ltd. 2020
J. R. Reagan and M. Singh, *Management 4.0*, Blockchain Technologies, https://doi.org/10.1007/978-981-15-6751-3

The drone software not only turned the steering wheel, turned the spray nozzles on and off, and raised and lowered the drill, but it also told the machines which route to take and where and when to turn. In its first year, the tractor first applied herbicide and then, over six hours, paused at programmed points to drill, plant seeds, and add fertilizer.

Drones equipped with infrared sensors and a robot "scout" monitored soil moisture and crop conditions, sending images to the researchers to view on their computers. Drones also collected crop samples for the researchers, who inspected them to determine whether they were ready for harvest.

"Throughout the year we've been predicting a yield of five tons. Looking in the trailer, it looks like we're not quite there. Our agronomist predicted 4.5 tons, and it looks like he's on the money."

"This project aimed to prove that there's no technological reason why a field can't be farmed without humans working the land directly now and we've done that. We set out to identify the opportunities for farming and to prove that it's possible to autonomously farm the land, and that's been the great success of the project."

"We achieved this on an impressively low budget compared to other projects looking at creating autonomous farming vehicles. The whole project cost less than £200k, funded by Precision Decisions and Innovate UK. We used machinery that was readily available for farmers to buy, open-source technology, and an autopilot from a drone for the navigation system."

Jonathan added: "Despite our combine being 25 years old, it performed absolutely wonderfully."

Challenges in those early years included adapting the drone autopilot technology for tractor use so that, rather than veering around an object such as a rock, it would travel in a straight line—essential to avoid damaging the plants in subsequent phases, such as during harvest. Quickly refitting the tractor for spraying after drilling and planting was difficult, as well, and monitoring the fields remotely was more difficult than checking them in person, team members said.

Yields that first year were a couple of metric tons less than a hectare likely would have produced if farmed using conventional methods, and time and money expenditures were much higher. But the potential benefits may outweigh the problems, researchers said: machines' ability to work when the weather is good, day or night; increased efficiencies, especially when several machines can work at once; better precision in farming practices, such as harvesting only the crops that are ripe; and a reduction in soil compaction.

The men told the British Broadcasting Company (BBC) that they see autonomous farming as the wave of the very near future.

"We believe … that in the future, farmers will manage fleets of smaller, autonomous vehicles. These will be able to go out and work in the fields, allowing the farmer to use their time more effectively and economically instead of having to drive up and down the fields," team member Jonathan Gill stated in a Hands-Free Hectare press release.

"Farmers will be able to concentrate on agronomic and business decisions while overseeing and managing a number of smaller automated machines, instead of sitting in a large tractor with 300-plus horsepower driving up and down a field," Martin Abell said in that release.

And the three men said they hope to see younger generations of farmers attracted to what has become, in the minds of many, an old person's job.

"It's looked down upon as dirty, mucky, hard work," researcher Kit Franklin said. "No, it's not: it's technical, it's advanced, it's exciting."

Case Study—Automotive

Daimler AG's Race to the AV Finish Line

German-based automotive company Daimler AG, maker of the luxury Mercedes-Benz automobile, anticipated Industry 4.0 more than thirty years ago when it launched its Prometheus driverless-car research project in 1986. Now the company is at the forefront of the autonomous-vehicle revolution in a number of markets.

One of the first to offer semi-autonomous features in its cars, Daimler made history in 2015 when it licensed two of its Freightliner Inspiration trucks for autonomous operation on highways in the US state of Nevada (In Europe, the truck is called the

Future Truck.). In 2018, it became the first international automaker to obtain a permit to test-drive its autonomous vehicles in Beijing, China.

Using cameras, sensors, artificial intelligence, "fast data" analytics enabled by supercomputing, the Internet of Things, GPS, and other technologies, Daimler was one of the first to make vehicles with "Level 1" and "Level 2" autonomous technologies, including:

- Remote calling, in which users summon their car via an app to pick them up,
- Hands-free driving along a pre-programmed route,
- Assisted lane-changing, triggered by the turn signal, changing lanes when sensors detect that it is safe to do so,
- Speed-limit assistance, reading speed-limit signs and using already available data to adjust the vehicle's speed,
- Brakes that halt the vehicle automatically if the driver stops operating it,
- Curve control, tilting the vehicle's body through curves for a more comfortable ride,
- Smart headlights that adjust themselves as needed,
- Pedestrian assistance, using lights to illuminate the path for crossing pedestrians,
- Blindspot assistance, using radar to detect vehicles in the driver's blind spot and, if the driver tries to turn in front of another vehicle, braking to bring the car back into its lane,
- Evasive steering assistance, stopping the car from spinning if the driver abruptly jerks the wheel,
- Emergency stopping, which warns with lights and noise of a possible crash before slamming on the brakes,
- Hearing protection, emitting a "white noise" that causes muscles in the ears to contract before a crash.

Daimler has also been among the leaders in autonomous trucking, seen as a solution for a shortage of human truck drivers, a way to increase efficiency and effectiveness of truck transportation, and a safer method of transporting people and goods by reducing driver fatigue and human error.

The automaker's semi-autonomous "platooning" system, for instance, has trucks traveling closely together at uniform speeds, reducing "drag" on the vehicles and saving fuel, as well as improving safety. Drivers sit behind the wheel, able to take charge at any moment, but the trucks drive themselves for the most part, communicating via computers and sensors to operate as a single unit—slowing down, braking, changing lanes, and speeding up simultaneously.

The company envisioned ramping up soon to "Level 4" status for its trucks, defined by the U.S. National Highway Traffic Safety Administration as being able to drive themselves in most instances, except during bad weather, or in construction zones or other situations requiring human control.

Daimler's third area of focus in the AV arena is driverless ride-hailing services. The company planned a pilot in the US state of California in 2019 in which AV cars carrying "human safety" drivers to monitor their systems would pick up drivers using an app and take them free of charge to their destination.

To develop and incorporate these technologies, Daimler opened an automated truck research center in Portland, Oregon, location of its US headquarters, and has spent hundreds of millions of dollars on research into automated driving, with an eye toward debuting Level 4 autonomous vehicles by 2025 at the latest.

The company has also entered into partnerships with high-tech company Nvidia to produce compact, high-speed supercomputers (processing 300 trillion operations every second as of this writing) for its vehicles with minimum power; and Bosch, one of the world's largest automotive suppliers, for the cameras and lidar laser, radar, and other sensors as well as software to direct the driving.

Autonomous Vehicles: The Five Levels of Automation (from the U.S. National Highway Safety Administration)

SOCIETY OF AUTOMOTIVE ENGINEERS (SAE) AUTOMATION LEVELS

0	1	2	3	4	5
No Automation	**Driver Assistance**	**Partial Automation**	**Conditional Automation**	**High Automation**	**Full Automation**
Zero autonomy; the driver performs all driving tasks.	Vehicle is controlled by the driver, but some driving assist features may be included in the vehicle design.	Vehicle has combined automated functions, like acceleration and steering, but the driver must remain engaged with the driving task and monitor the environment at all times.	Driver is a necessity, but is not required to monitor the environment. The driver must be ready to take control of the vehicle at all times with notice.	The vehicle is capable of performing all driving functions under certain conditions. The driver may have the option to control the vehicle.	The vehicle is capable of performing all driving functions under all conditions. The driver may have the option to control the vehicle.

Case Study—Consumer

Alibaba

After its founding in 1999, Alibaba grew to become the world's number-one retailer and China's largest e-tailer, controlling 80% of that country's online market.

Not only did it build a formidable online presence with marketplace sites selling to businesses and consumers—and enabling them to sell to one another—but it also offers shopping search engines, cloud computing services, and Alipay, the most popular online payment system in China.

In 2017, however, the retail giant began transforming itself again with an omnichannel model called "New Retail." Blurring the boundaries between online and offline, or so-called bricks-and-mortar, sales, New Retail focuses on providing customized, seamless shopping experiences from phone to tablet to computer to store to home, integrating online and offline shopping (*O2O*).

Alibaba's New Retail uses digital technologies, including artificial intelligence, cloud computing, virtual reality, and data analytics, to provide an integrated shopping experience tailored specifically for each customer. To do so, it had to recognize that not everyone wants to shop online—at least, not all the time. For high-ticket items

© The Editor(s) (if applicable) and The Author(s), under exclusive license to Springer Nature Singapore Pte Ltd. 2020
J. R. Reagan and M. Singh, *Management 4.0*, Blockchain Technologies,
https://doi.org/10.1007/978-981-15-6751-3

such as cars and kitchen appliances as well as personal items such as cosmetics and groceries, many still prefer the experience of "try before you buy."

To answer that need, Alibaba began acquiring partners along the value chain, including:

- Bailian Group, the largest department store chain in China. The two companies agreed to develop new retail technologies, including artificial intelligence, the Internet of Things, and big data, share customer data, and combine their memberships using geolocation, facial recognition, and other systems. Shoppers in Bailian stores may use Alipay, owned by Alibaba affiliate Ant Financial, which has incorporated Bailian's payment cards into its system as well. Alibaba also benefits from Bailian's supply chain management and logistics.
- High-end retailer Yintai Group, owner of Intime Retail, is a leading Chinese department store and mall operator. After becoming Yintai's chief shareholder, it began exchanging goods with Intime to sell in stores and online and sharing technologies, including big data.
- Electronics retailer Suning, combining Alibaba's online assets and Suning's physical stores and distribution.
- Sanjiang Shopping Club, a regional supermarket chain, which provides Alibaba with access to its retail network and consumer data.
- Sun Art Retail Group, China's top grocer at the time of the partnership, running RT-Mart and Auchan supermarket-department ("hypermarket") stores. The chain, which became popular for customizing its product offerings to local needs—placing hot sauces on shelves next to clams, for instance—had seen its sales slow down as customers moved their purchases online.
- Logistics platform, Cainiao, uses a combination of ships, trucks, and bikes to transport millions of packages per day. Becoming its major shareholder took Alibaba closer to its goal of strengthening and expanding its global delivery network while cutting shipping costs and reducing the risks of relying on third-party shippers.
- Fresh-food chain HEMA, with its curated inventory of 3000 products from 100 countries as well as catered and fine dining offerings. The partnership allows customers to order online via a mobile app and receive free delivery within half an hour in a 3 km radius—a feature made possible by Cainiao—or scan barcodes at the store, pay using the app—linked to Alipay—and set up delivery for a true omnichannel experience. The app captures customer shopping behavior all along the way, allowing Alibaba to personalize user experiences with custom promotions and tailored suggestions.

Case Study—Energy

San Diego Glass & Electric

In 2009, San Diego Gas & Electric (SDG&E) in the US state of California became one of the first utilities in the world to use "smart" technologies when it installed 2.3 million digitally connected natural gas and electric meters in homes and businesses throughout its service area, at a cost of $1.042 million.

The Technologies

To install, service, and maintain its smart grid, SDG&E entered into partnerships with many providers for metering, communications, transmission, distribution, consumer engagement, and other services. Technologies include:

- High-speed wireless systems to communicate with smart devices on transmission and distribution poles,

J. R. Reagan and M. Singh, *Management 4.0*, Blockchain Technologies,
https://doi.org/10.1007/978-981-15-6751-3

- Updates and expansions to supervisory control and data acquisition (SCADA) capabilities,
- Fiber for remote communications, monitoring, and control of transmission and distribution equipment,
- Upgraded communications at substations and supporting telecom sites,
- *Synchrophasors*, which provide a real-time measurement of electrical quantities across the power system,
- A *low-power communications network*, a wireless radio system that provides low-speed, low-power, wide-area communications for remote monitoring of overhead and underground fault circuit indicators, smart transformers, Federal Aviation Administration tower obstruction lights, and other low-bandwidth equipment,
- Unmanned Aircraft Systems, or drones, to inspect power lines,
- Home area network devices such as digital thermostats, and the applications to use them.

The Benefits

SDG&E has seen many rewards from using connected technologies. Published case studies, and the utility's own reports, indicate that "going smart" has changed virtually every aspect of its operations.

Benefits of the digitization include:

- **Cost savings**: Meter read themselves and send usage data to SDG&E's billing department, saving the cost of sending personnel and trucks through its 4100-square-mile service area. Also, the utility no longer needs to schedule and send personnel to turn the power on or off but can do so remotely. SDG&E expected to save $369–615 million in fuel costs alone through 2020.
- **Energy storage**: Customers can save money by charging electric vehicle (EV) and storage batteries during low-demand, lower-rate times, and using the power later, when demand and rates go up. SDG&E also stores energy in its lithium-ion battery facility, one of the world's largest.
- **Sustainability**: An ever-increasing number of customer-installed rooftop solar systems provides power to the utility to help it comply with California laws and meet energy needs during peak demand. More than 1000 MW of solar and wind energy flow on peak days to SDG&E customers from more than 114,000 power producers, mostly solar.
- **Load management**: SDG&E's "Reduce Your Use" demand-response program offers bill credits to customers who turn down the power on peak-load days, avoiding overloads to the distribution system. Otherwise, the utility might have to build costly new power plants or transformers to ensure power delivery on those few high-demand days per year.
- **Outage management**: Instead of relying on customers to call SDG&E when power is out, utility personnel now hear from the meters—sometimes before outages even occur. The utility can determine the extent of an outage and even

its cause and may be able to restore power remotely without sending anyone to the site. The average customer now experiences one 60-min power outage every other year.

- **Better decisions**: Smart data and analysis tell utility leaders where power is coming from and where it's going, where demand is shifting, and where existing transformers and other equipment need replacing or augmenting.
- **Reliability**: Looming storms, impending heat waves, and even fog can alter the delicate balance of power generation and distribution needed to reliably manage the grid. Incorporating weather-system data with meter data allows SDG&E to see where power needs to increase and for how long, and to anticipate when to turn it down to accommodate energy coming from solar and wind sources. Power flows more smoothly, without surges that can damage delicate computer circuits.
- **Proactive problem-solving**: Smart technology tells SDG&E when the voltage entering a customer's home or business is too low or too high and can even spot internal wiring issues. The data meters generate show when they've been tampered with for energy theft or other reasons. Smart devices also alert when transformer loads approach the "red" zone, enabling the utility to replace or repair transformers before they shut down, avoiding outages.

Challenges and Lessons Learned

Costs: Recovering the costs of designing, deploying, and maintaining a smart energy grid were high—estimated at more than $1 billion from 2009 through 2020—but the utility estimated social, environmental, financial, and other benefits of at least $3.1 billion.

Acceptance: Some customers resisted having smart meters installed. Privacy and health concerns were the main reasons why. Anticipating controversy, SDG&E proactively sent letters to its customers about the changes and held meetings to answer questions. Ultimately, fewer than 1% of customers initially opted out of the new system.

Communication glitches: Designing a network with enough devices to effectively and smoothly communicate was a challenge at first, especially in low-population areas.

Customer engagement: For customers to make use of the savings and other features smart technology offers, the utility had to make sure its systems and applications were user-friendly.

Changing company culture: Employees—even on the business side—had to develop not only new, tech-savvy skillsets but also new mindsets, as the utility shifted from a paternalistic, "we-know-best" power provider to one working in tandem with customers. To make these culture changes, SDG&E established a "smart grid team" of employees from across the company to educate co-workers and get buy-in to the new, customer-centric mentality.

Case Study—Environment

South Korean Smart Cities

In 2008, global financial crisis makes slowdown that not makes tougher for people livelihoods in many countries but also need to the targets set to fight global warming in upcoming years. Due to economies needed cash infusions to get back on track, investments in global climate change solutions have been postponed. However, some of countries such as South Korea, Singapore, Switzerland, etc., have decided to use billions of euros to invest in clean energy projects. South Korea has been lead the development with spent 80% of its €30.2 billion euro. Additional, South Korean government has planned to spent 2% of it GDP will be spent over coming five years.

© The Editor(s) (if applicable) and The Author(s), under exclusive license
to Springer Nature Singapore Pte Ltd. 2020
J. R. Reagan and M. Singh, *Management 4.0*, Blockchain Technologies,
https://doi.org/10.1007/978-981-15-6751-3

The South Korea has been taken action to reduce the level of carbon and go green environment in their cities. As of now, South Korea imports 97% of its energy so that it's need larger commitment to "green growth" and the creation of green technologies with energy security. South Korean government searching a way to minimize the imports energy and investing heavily in renewable energy sources.

The South Korea smart cities projects are directly linked with Korean government national strategy of Green Growth and also provide the sufficient incentives for international companies to make real smart city concept. This appendix aims to focus on development of smart city projects in South Korea, where Korean government is focusing to the development of New Songdo, Saemangeum, and Sejong City.

Smart City Projects in South Korea

In recent years, the implementation of the 2006 U-Korea Master Plan has begun, a strategy for ubiquitous growth that will help to address many of the problems created by the high urban density that characterizes South Korea.

Such proposals have been adopted by the government at all levels. The central and local government authorities found sustainable growth of U-cities in the construction of electronic city (ICT) infrastructure and service providers.

The plan gave some direction to the urban planning process in Korea (especially e-city planning) as well as reflected the increasing importance of green cities. In 2008, the Korean government passed the "Ubiquitous Construction Act." The Act aims to facilitate multi-sectoral convergence, e.g., construction and ICT; pan-government planning for big cities, i.e., top-level planning from central government to local government and IT. In addition, the National Information Society Agency (NISA) has established a U-city Infrastructure Guideline to help prevent expensive duplication of technology at the local level.

Local Authorities (Metro Cities) have started their own smart city plan all over the country. The Seoul Metropolitan Council recently completed a major plan that is the Digital Media City (DMC) within Seoul. The DMC will plan to be make worldwide IT industrial hub in Seoul. It is a high-tech complex based on emerging technologies such as television, video, gaming, music, e-learning, and related industries. The DMC consists of a cluster of business offices, residences, galleries, conference halls, and cultural centers.

The Korean government has started and sponsored to develop the U-City with the help of the ministry of Land, Transport, and Maritime Affairs (MLTMA), the Korea Institute of Construction and Transportation Technical Evaluation and Planning (KICTTEP), the Academy and other Agencies. Below has mentioned some of project names as example:

• The Mega Korean Construction Technology Project:

   – Aim: To raise the quality of life of people,
   – Total budget: €79.2 million.

- The Korean Land Specialization Program

  - This project is based on PPP model (public and private partnership),
  - Total budget: €96.7 million government funding, €33.3 million by private funding.

- The establishment of a Center for Sustainable Housing

  - Aim: Order to cut on the carbon footprint by 40% through the use of technology.
  - The establishment of the Construction Waste Recycling Research Center.

Incheon Free Economic Zone (IFEZ)

A similar cluster of developments has been undertaken around Seoul, the most ambitious of which is the €101.5 billion development of the Incheon Free Economic Zone (IFEZ), located 1 hour southwest of Seoul. This includes Songdo International Business City, Cheongna Leisure City, and Yeongjong Global Logistics City.

The construction of Songdo International Business City is approaching its final phase, with most of the infrastructure already in place. Approximately 27,000 residents will be displaced by the end of the year, and with the end of the project planned for 2016, the authorities expect the city to accommodate 65,000 residents in this smart city, with an additional 300,000 on a daily basis. This is the biggest private real estate investment ever and is estimated to cost about €27.8 billion. It is built on the peninsula off the coast of Seoul and hosts offices, houses, shops, hotels and public areas.

Cisco, a large American IT company, has been tasked with connecting the city with optical fiber to the various systems that keep Songdo operating. TelePresence will be built in homes, workplaces, hospitals, and shopping malls so that people can make video calls anywhere they want to. Sensors are deployed in streets and buildings to regulate everything including temperature to road conditions, enabling the city to operate properly and respond to problems quickly. These scanners also monitor things like fire and safety in the many towers and control the amount and quality of the water in Songdo Central Park. RFID car tags will help to alleviate congestion by handling road traffic; traffic lights will use very powerful LEDs. The energy consumption of homes and their electrical appliances will be monitored automatically to better understand how residents use energy and adapt the grid to make supply and demand more efficient. In addition, around 40% of Songdo's area will be green, including rooftop vegetation to help cool the city. Rainwater reservoirs and "black water" recycling from sinks and dishwashers would further minimize the need for fresh water. Waste trucks will also be a thing of the past with a pneumatic solid waste management system that sucks wet and dry garbage straight to the dump through a piping system.

Yeongjong Logistics City, a second part of IFEZ, includes an industrial and logistics complex offering a range of exhibition spaces, accommodation, accommodation, business and education. The large seaport and the large airport of Incheon (six times the best airport in the world) are also situated in the vicinity.

Milano Design City in Yeongjong is among the landmark projects and is expected to become a network of cultural, architecture and educational facilities by 2017. Determined according to the method of Cooperation between Milan and Incheon, which was established in November 2008, the Milano Design City project is expected to employ 35,000 people.

The Medi-City, also built in Yeongjong, will include international hospitals, hotels, and research facilities to provide a wide range of medical services, new product production, education, entertainment, and leisure.

Cheongna Leisure City is at an earlier stage of development. Various projects are now being created, including the International BIT Port Project, the GM-Daewoo R&D Center, the Golf Course Theme Project and the Incheon High-Tech Park Project. The global BIT Port project is a collaboration venture between the Seoul National University (SNU) and the Korea Advanced Institute of Science and Technology (KAIST) to establish a Biology Technology and Information Technology Education and Research Cluster. In addition to stimulating development and investment allocation in Cheongna, it will provide a strong foundation for IFEZ's green growth.

In the meantime, the GM-Daewoo R&D Center is a large-scale initiative that will make a major contribution to the economic development not only of Incheon but of Korea as a whole. In line with IFEZ's strategic goal of being the most innovative city in the world via the use of ubiquitous and wireless technology, Incheon High-Tech Park can further draw international business to Korea as one of the most technologically sophisticated cities in the world.

Saemangeum Development Project

On February 1, 2010, the South Korean authorities presented a blueprint for the transformation of Saemangeum, a large reclaimed tidal flat located in North Jeolla Province, into the economic hub of Northeast Asia. In 1991, the government began the Saemangeum program with the original goal of creating farmland by reclaiming the tidal flat along the southwestern coast. The government revised the initial farmland program as the nation no longer needed additional farmland because of oversupply of rice and the declining agricultural sector. In 2008, the government cut the farmland ratio to be built from the original 72–30%. In addition to agriculture, the emphasis will be on industrialization, recreation, research initiatives, the environment and renewable energy and the construction of a new city. For this purpose, eight different complexes will be constructed:

- Agricultural complex,
- Industrial complex,

- Tourism and leisure complex,
- International business complex,
- Science and research complex,
- New and renewable energy complex
- Residential complex,
- Ecological and environmental complex.

In April 2010, the Dutch Ministry of Economic Affairs, Agriculture and Innovation signed an agreement with the Korean Prime Minister's Office on the sustainable development of the Saemangeum region. The focus of this MoU is on the sustainable cooperation and sharing of information and technology between the two countries in the water, sustainable agriculture, renewable energies, dredging, design and architecture, ecology and recreation. Annual workshops on sustainable agriculture are already taking place and incentives are being actively sought by Dutch companies and organizations, sponsored by the Dutch Embassy in Seoul.

The whole project is very ambitious and has a long horizon for implementation. The plan currently has set the project completion for 2020 so that there is ample opportunity for Dutch companies to enter into contracts that will be tendered on time. The key problem remains, however, whether the South Korean government would be able to find adequate funding and expenditure to finance this initiative. There has been a shortage of funding so far, and while the Chaebols have also contributed to the initiative, there is still a significant investment gap. So the success of this project relies on how well the South Korean government will be able to market the project and how the economy will grow in the near future.

Sejong City

Another ambitious project that is worth mentioning that has been almost completed is the construction of the brand-new city of Sejong. This city is meant to become the new administrative capital of South Korea, housing 36 ministries and government agencies and more than 10,000 civil servants. The idea is to create a more efficient government center.

The building of this city is entering its final phase, but the area is still more or less quiet, with no movement to be found in its streets. Let us hope that this will improve in the coming months; it seems that the development of a new city is always followed by an initially slow rise in population. Government incentives should change this, but there is still a danger that people will only work in this city during the weekdays and return to their families in Seoul on the weekends.

Smart Grid Initiatives

South Korea has been at the frontline of designing smart grid systems, and its priorities are very ambitious. The drive for smart grids stems from a wider emphasis on the CO_2 emission goals that the South Korean government has set itself. They plan to reduce CO_2 rates by 30% from business as normal (BAU) by 2020. This can be considered as ambitious target, because the energy-intensive industry (steel, semiconductor, cement) has been rising at a steep pace in recent decades. The Roadmap of Smart Grid Initiatives focuses on five sectors:

Smart Power Grid

The goal here is to innovate in the interconnection between consumption and supply sources. This will allow new business models to emerge and improve power grid failure and automated recovery systems to ensure stable and high-quality power supply.

Smart Consumer

Having real-time information available to consumers will encourage them to save energy and to produce smart home appliances that adjust their operating powers in line with energy prices will contribute to efficiency.

Smart Transportation

The goal is to develop a national charging network that will enable electric vehicles to be charged anywhere. A vehicle-to-grid network is also designed to charge the batteries of electric vehicles during off-peak hours and to resell excess energy during peak hours.

Smart Renewable Energy

The ultimate aim here is energy self-sufficiency of homes, schools, and villages. The goal is to develop smart renewable energy generation complexes across the country by rolling out microgrids and installing small-scale renewable energy generation units for every end-user.

Smart Electricity Service

Expanding the right of choice of customers is the goal here, by expanding the selection of energy-saving electricity tariff plans. Bringing ICT and electricity closer together will also benefit consumers by adding new electricity services. Real-time electricity trading systems for energy and derivatives transactions are also planned.

In order to make tangible progress on this dream, South Korea has launched a smart grid pilot project on Jeju Island to explore the feasibility and effect of a smart grid. So far, 6000 households have been connected to the smart grid project, which is expected to transform the way power is delivered to our households.

This pilot project is the first phase of the deployment of the Roadmap (technical validation). The second level, which is scheduled for 2012–2020, is aimed at extending this program to include metropolitan areas (intelligent consumers). The final step, expected to be completed between 2021 and 2030, is aimed at the completion of a national power grid (intelligent power grid). Each of these phases provides some opportunities for foreign companies.

Case Study—Finance Technology

Manulife/John Hancock

The Manulife Financial Corporation, with headquarters in Toronto, Canada, is a 130-year-old insurance and wealth management company with additional offices in the USA (where it operates as John Hancock Financial Services, Inc.) and Asia, and annual revenues of $46.5 billion.

Today, the organization is setting itself apart in the very conservative financial sector with a digital transformation project aimed at providing its twenty-first-century customers with up-to-date services, and positioning itself as a digital innovation leader. To get there, the enterprise has focused first on people, then on processes, and then, at last, on technologies.

Culture Change from the Top Down

People

It's a truism in the business world that change in organizational culture begins with the people at the top. In late 2016, in its initial foray into digital transformation, John Hancock recruited as its senior vice president a chief marketing officer with more than 20 years' experience. The following year, Manulife hired a CEO whose vision included a "technology first" approach.

These technology-savvy leaders were well positioned to inspire change in the organization's internal operations and business culture. They worked with managers to establish a broad agenda aiming for more, better, and faster innovations and collaborative processes that gave employees more ownership and empowerment.

J. R. Reagan and M. Singh, *Management 4.0*, Blockchain Technologies, https://doi.org/10.1007/978-981-15-6751-3

Processes

To keep up with the fast pace of change in the fourth industrial revolution, Manulife/John Hancock established "innovation teams" bringing together workers from various departments to brainstorm new ideas. They gave these teams free rein to work outside corporate bureaucratic restraints—as though they were startup companies—so their innovations could move more quickly to market.

For instance, the innovation group taking charge of Twine, a mobile investment app, not only worked together on its development and launch—with all having an equal say—but then shared its experiences and lessons learned so other teams could benefit.

Ultimately, the entrepreneurial-collaborative-small-team approach would infiltrate the entire organization, changing processes all the way to its core.

For example, Manulife launched an "advanced analytics" group made up of data scientists, data engineers, and strategists from its markets around the world to focus on data analytics and marketing; talent analytics; *underwriting*, or the process of deciding whom to give insurance coverage to, the kind of coverage they should get, and what to charge; and fraud detection.

Technologies

To increase its digital capabilities, Manulife has partnered with a number of technology companies. One developed an artificial intelligence tool to help portfolio managers make better decisions. Another, Vitality, created an app enabling customers to share their fitness data in exchange for insurance discounts. The data helps Manulife/John Hancock better serve its customers by offering incentives and push notifications promoting healthier lifestyles, and enables more targeted, customized products and services to fit individual client needs.

Ultimately, Manulife/John Hancock envisioned using one platform to provide its customers with a number of apps including Vitality and Twine as well as other digital offerings. To get there, it partnered with a global consulting and technology firm. The organization hoped to save money with the consolidation and improve customer service as well as its own flexibility to innovate, engineer new features, and incorporate next-generation technologies including the Internet of Things, AI, and blockchain.

Case Study—Manufacturing

Adidas's "Speedfactory" Plants

Until a few years ago, the athletic-wear manufacturer Adidas produced its shoes and other sporting gear in factories located primarily in Southeast Asia, then shipped its products to retail outlets around the world—a process that took many months and entailed many challenges.

Producing in cheaper labor markets distant from corporate headquarters meant that the company had to split its operations. While its in-house staff designed and developed products and handled marketing and sales, factory laborers on the other side of the globe stitched, glued, and laced Adidas's shoes and other sportswear. The time from design to production for a new style averaged 18 months.

Also, exercising quality control was difficult across the miles, and scaling was a monumental task: Sprawling factory complexes only made shoes in lots of 20,000 per shoe size. Distribution was slow, as well, with shipping from factories to retail outlets taking as long as six weeks.

J. R. Reagan and M. Singh, *Management 4.0*, Blockchain Technologies, https://doi.org/10.1007/978-981-15-6751-3

Entering the Digital Age

Today, Adidas is using automation and digitization to resolve these and other challenges. "Speedfactories" can produce a single, custom-made pair of sneakers in as little as one day after a customer orders it, using only a small human workforce. Because the factories are near their target markets, shipments arrive soon afterward. Like the shoes they produce, these factories are designed for speed—enabled by connected technologies including data, virtual reality, robotics, AI, and 3D printing.

The journey from conventional manufacturing to Speedfactory has been a long and gradual one, catalyzed by consumer demand and competition. Although Adidas has been in existence since 1949, the company, like so many others, grappled with how to conform to the new paradigms established by the Internet.

Keeping up with consumer demands, for instance, became all but impossible. Shoppers were no longer content to wait to wear the styles they were seeing on social media, but with an 18-month lag between the start of design and production, Adidas consistently fell behind. Without speed to market and a nimbler business model, the company knew it risked irrelevance.

Examining what others were doing, the company noted that Apple and Amazon were collecting and analyzing data to improve their products and processes. In 2014, Adidas analyzed its own data to initiative using trend analytics and predictions, to little avail—it could not keep up with fast-changing consumer demands. The company tried speeding up the supply chain, aiming to reduce wasted time and speed up orders, but could only shave days or weeks off the process.

Adidas's leadership noted that "fast fashion" companies had built smaller factories near markets to move items from sketch to store in just a few weeks. They also recognized that what Adidas offered wasn't that different from their competitors' selections because all were choosing components from the same suppliers. If the company wanted to stand apart from the competition, it would need to reinvent itself to emphasize innovation, not replication.

And so the company found new materials including a springy, lightweight plastic that it developed for use in its popular "Boost" shoe, introduced in 2013, which revitalized the Adidas brand. To meet consumers on their own turf—the digital field—and keep up with their demands by becoming flexible and fast, Adidas opened its first Speedfactory in 2015.

Located in Ansbach, Germany, near Adidas's corporate headquarters, the smaller, mostly automated plant used digital designs that could be altered according to even a single consumer's needs and desires and sent to robotic arms and 3D printers for precise custom fabrication. To design and build the plant, the company partnered with a manufacturer of precision parts and components, as well as a robot manufacturer and a technology company.

Almost every step of shoemaking at Speedfactory happens digitally. Trainers get designed and their virtual prototypes tested on a computer screen, and the manufacturing process gets simulated digitally, as well. The foot size, gait, and other qualities of the customer may be scanned and entered into the system, too, for a more tailored shoe.

Then, the Speedfactory makes most of the shoe components from raw materials such as plastics and other synthetic materials and fibers—robotic machines knit the uppers and 3D printers produce the midsoles—and welds them together using lasers.

To finish the sneakers, however, humans must perform the final step in the production process, doing a task that machines still cannot: lacing the shoes.

Boost made Manz look afresh at the way Adidas made shoes. The robotic knitting system named Primeknit [1] Crucially, both methods not only created new products, but also incorporated previously different operations. Whereas sports shoes were typically made from a lot of individual parts, Primeknit knitted the entire top in one seamless whole. Manufacturing and installation were parallel.

At the same time, Manz started experimenting with the artificially produced fiber "Spider silk" manufacturer AMSilk [2] that could be tuned to consist of different properties on demand. He partnered up with Carbon, a Google- and General Electric-funded company with a new form of 3D printing that used ultraviolet light to correct the shape of the template after it had been taken out of a vat of liquid resin like Excalibur. 3D printing has always been too sluggish, costly, and heavy to use in shoes. It seemed now that it could finally be ready.

It's Friday afternoon in the Speedfactory, and the week's last batch of Made for London is almost finished. The soles are molded, the uppers knitted, patched, and stitched. Now the parts just need to be put together.

In a factory in China, this would be an unpleasant task. To assemble shoes, workers there use glue. Wearing masks to protect themselves from the fumes, they apply several layers, then press the parts together and wait for it to dry. In the Speedfactory, that process is eliminated—because the Made for London is not glued but welded by lasers.

Seen in action, the welder resembles a high-tech blacksmith's forge. In a glass-fronted cabinet the huge size of a fridge, a yellow light surrounds a pair of upside-down trainers. The light fades, then glows again, brighter and red with heat. As the parts are effectively melted together, the smell of burning rubber fills the air. Then a burly, bearded man in a sweaty T-shirt opens the doors, takes out the shoes and sets them on a ledge. The team gather around to admire their creation. "In China, it's an imprecise process," says Carnes. "This makes sure everything is lined up before they touch it."

Customization

Eventually, Steyaert says customers may be able to create their own completely custom, one-of-a-kind Adidas shoes designed to their own specifications online. Things like patch placement and details on the uppers of shoes would be able to be customized.

"Speedfactory is able to customize the shoe indefinitely while being in an automated engineering process," Steyaert said. "We can actually tune the shoe to the customization that the consumer wants to have. That's the goal: full customization, but without compromise on speed."

Data Collection

Digitizing production brings many benefits such as quality control. Another is planning: If Manz's team want to prioritize an incoming order of Stan Smiths, or change the plastic on a shoe upper, they can simulate the process using the Speedfactory's "digital twin." "It's a factory in the web," says Manz. "Traditionally, we would just set [the production line] up and see what happens," says Steindorf. "With the digital twin you can simulate consequences."

Human Labor

Improvements in automation can now finally substitute for cheap foreign labor, which will naturally push factories closer to where the consumers are. As manufacturing shifts from offshore mass production to customized, local fabrication, new jobs will open up for human workers, some of which have yet to reveal themselves. "We used to have distribution built around manufacturing," Mandel says, referencing the centrality of offshore factories, "and now I think that manufacturing is going to be built around distribution."

At every stage, the Speedfactory is both more and less than it seems. Take automation: There are plenty of robots, sure, but also a good deal of old-fashioned manual labor. Next to the patching conveyor belt, a line of middle-aged women stitch uppers at sewing machines. After that, a man steams the parts by hand.

"To be honest, it's difficult to find those kinds of workers now because this industry barely exists in Europe," says Steindorf, who oversaw the hiring. Adverts for experienced stitchers turned up women who'd worked in shoe production before it moved to Asia 30 years previously (To refresh their skills, Oechsler brought in a cobbler who still practiced traditional shoemaking.). But other positions couldn't be filled in the same way, for the simple reason that the work involved wasn't done anywhere else in the world. Workers had to learn their jobs from scratch.

1. (http://www.wired.co.uk/article/adidas-samba-primeknit).
2. (http://www.wired.co.uk/article/adidas-futurecraft-biofabric-shoes).

Case Study—Media and Entertainment

Media and Entertainment: Google VR

Google has long been known for its Web-based services including its ubiquitous search engine. But the company has invested in a number of emerging-technology initiatives, as well, including virtual reality. For years, its efforts yielded little fruit, its engineers struggling with the same challenge other VR designers faced, of streamlining the clunky, expensive hardware.

Out of Google's "20 percent project" innovation program, in which employees devote one day of their work week to developing new ideas, came a solution to these obstacles. Two Google staffers wrapped a smartphone in cardboard and cut holes through which to view the screen, and presented it as a simple, lightweight, inexpensive way to present VR. Google Cardboard debuted in 2014 for $15, and by early 2017 the company had shipped 10 million of the devices and offered the design online so anyone could make their own for free. After years of research and development, VR had finally gone mainstream.

© The Editor(s) (if applicable) and The Author(s), under exclusive license
to Springer Nature Singapore Pte Ltd. 2020
J. R. Reagan and M. Singh, *Management 4.0*, Blockchain Technologies,
https://doi.org/10.1007/978-981-15-6751-3

Today, VR and its cousin, augmented reality, are hailed as the wave of the future not only for technology but also in media and entertainment—and Google continues to lead the way with low-cost hardware and a growing catalogue of software and programs.

The Technologies

VR and AR rely on a number of technologies, including:

- *Cameras* to capture the images used in immersive, 360-degree scenarios,
- *Microchip sensors* to track user movements, including *accelerometers* to detect three-dimensional movement (such as walking or throwing something), *gyroscopes,* to detect angular movement, such as looking up or down, and *magnetometers*, a form of compass, to track the direction the user is facing,
- *Processing hardware and software* to manage devices, analyze incoming data, and generate the proper user response in near-real-time,
- *Positional audio,* using multiple speakers to provide the illusion of the VR environs' surrounding the user as well as to provide information,
- *Video graphics* to augment real-life scenarios and to create immersive experiences,
- *Artificial intelligence* to enable conversations and enhance interaction in the virtual/augmented realm.

The Benefits

Virtual and augmented reality technologies can enhance many experiences of life and enable users to have new ones—walking on the moon, flying an airplane, diving deep under the sea—that they might never enjoy otherwise. Benefits include:

- **Affordable "travel."** Google Earth VR takes users to locales around the world such as New York City in America, the Great Pyramids of Giza in Egypt, and the ruins at Machu Picchu in Peru.
- **Education**. Google Expeditions enables educators to take students on virtual fieldtrips to explore in 3D the lessons they learn in class—from dinosaurs to Renaissance sculptures.
- **Safe adventures**. Google Cardboard apps offer the chance to explore the solar system, climb mountain peaks, jump off skyscrapers while doing stunts, and more without danger or risks of physical harm.
- **Interactive entertainment**. Enjoy the game courtside or watch from the sidelines, without leaving your home; immerse yourself in Netflix and YouTube videos.
- **Social awareness**. Tour refugee camps and experience the devastation caused by natural disasters and learn first-hand about poverty and other social challenges.
- **Accessibility**. People with disabilities can use VR to have experiences that their physical limitations prevent.
- **Social connections**. VR can bring people together from around the world in a variety of situations to connect, share, play games, and have experiences together.

Challenges and Lessons Learned

Google Cardboard and its slightly-higher-end sibling, Google Daydream, might have solved the problems of ease of use and affordability for VR and AR technology, but obstacles remain to the company's dominating the VR market:

- **Fewer features.** "You get what you pay for," the saying goes, and it seems to be true with VR technology, at least in the early stages. Google's less-expensive devices may serve as an entry point for more casual users—as opposed to use-heavy VR gamers—but as costs come down across the line, the company could find itself at a disadvantage compared to the heavyweights, which offer higher-quality imagery and more realistic and interactive features.
- **Quality of programs.** One common complaint is a paucity of high-quality VR content. Google and other technology companies are working to address this issue by offering open-source software to developers free of charge.
- **Interoperability.** Google initially designed its Daydream headsets specifically for "Daydream-ready" phones made by its own and partnering companies—which do not include Apple. This proprietary approach limiting Google's penetration in the VR market. And it isn't the only VR developer working in a silo. The technology is still nascent, and a variety of players are developing tracking, input, and content technologies without standardization to guide them. This "console wars" situation is common with emerging technologies, and Google and others in the VR sector will likely learn to coordinate even as they compete.

Case Study—Medical

Medical Wearables

From the fifteenth century, when the artist Leonardo da Vinci invented the *pedometer,* an instrument that counts its wearer's steps, humans have envisioned using machines to help improve fitness and health. Connected digital technologies have expanded the concept. Wearables today not only measure and document physical activity, but also monitor and even diagnose a growing number of medical conditions; alert wearers when they need to seek medical attention and send information to healthcare providers. They also help caregivers monitor hospital patients. Using a combination of technologies, watches, bracelets, phone apps, ingestibles, patches, clothing, and more holds the promise to lengthen, strengthen, and improve quality of life.

The Technologies

Medical wearables come in many forms, but all work in a similar fashion: They measure the body's responses, vital signs, and movements, and analyze and report the results to their wearers as well as, when authorized, remotely to medical professionals.

The *Internet of Medical Things* uses technologies including:

- *Sensors*, which attach to the body or breath to measure heart rate, activity levels, brain waves, perspiration, body temperature, and other indicators of health,
- *Cameras,* to help visually impaired people "see" as they move about, and to provide navigational information,
- *Data analysis*, to report results of monitoring and tracking in meaningful and actionable formats,
- *Bluetooth*, to communicate data results to the user's smartphone or other computing device,

J. R. Reagan and M. Singh, *Management 4.0*, Blockchain Technologies, https://doi.org/10.1007/978-981-15-6751-3

- *Wi-Fi*, to send data to the cloud or other remote location for processing, analysis, sharing, and storage,
- *Cloud technology*, for remote, anytime access to information,
- *Artificial intelligence*, for using data and context to diagnose illness and disease.

The Benefits

Medical wearables stand to empower people as never before to improve their health by enabling earlier diagnostics of medical conditions and providing alerts about possible health issues as well as guidance for nutrition, activities, medications, and more.

Uses and potential uses of the technology include:

- **Activity tracking**. Instead of the old-fashioned pedometer, a mechanical device that worked like a vehicle's odometer to advance measurements step by step, sensors in smartphones, smart watches, and other devices track the number of steps taken, distance traveled, and other fitness activity metrics.
- **Continuous glucose monitoring**. Without drawing blood, devices including watches and skin patches monitor blood glucose levels in diabetes patients, and issue alerts when numbers go askew.
- **Blood pressure and heart rate monitoring**. Skin patches using ultrasound can monitor blood pressure in deep arteries continuously and in *real time*, or as it happens, especially valuable during surgery and for patients with heart and lung problems.
- **Fall detection**. For senior citizens with brittle bones, a fall can be deadly—but non-seniors, too, benefit from wearables that sense when they fall, and offer to call emergency services.
- **Navigation**. Cameras, sensors, and artificial intelligence combine to help people with visual impairments find their way from one place to another.
- **Patient identification and records**. Hundreds of thousands of hospital deaths every year result from patient misidentification. Wearables such as wristbands with RFID tags can provide caregivers with not only the identification of the wearer but also other important information such as blood type, allergies and sensitivities, and medical history.

- **Future uses**. Many health-related uses for wearables are in the making, including stress management, body-temperature adjustments, and head injury detection for athletes during sporting events.

Challenges and Lessons Learned

Accuracy: With wearers' health and possibly their lives at stake, the accuracy of medical wearables' data is critical. With many different sensor types and manufacturers from which to choose, assessing their quality is important as well as certifying that the wearables using them work as they should.

Interoperability: The data sensors collect is useless unless it can be shared with and used by other apps and devices. A fitness tracker must be able to communicate with a cardio device and a calorie counter, for instance, to provide the patient and caregiver with an overall prognosis and tailored treatment plan.

Privacy: Our physical, mental, and psychological medical information is among the most sensitive and personal information available about us. The data medical wearables collect, analyze, share, and store must be protected from unauthorized eyes, whether in the cloud or in the devices used to process it.

Case Study—Retail

"Alexa is Everywhere"—Amazon Alexa

When Amazon introduced its Alexa-enabled Echo voice assistant in 2015 in the USA, the device was the first of its kind—and met with derision, with some calling it useless and others decrying it as an invasion of privacy. Within just a few years, however, Alexa would incorporate into more than 20 million devices in more than 80 countries.

Amazon's vision of a smart speaker has grown and changed since the Echo's incipience. Today, the device does so much more than play music; it also orders pizza, helps with online shopping, calls for a ride, tells users about their day, answers questions, makes dinner reservations, turns lights on and off, and so much more—in part because Amazon has made its voice-assisted technology available to other developers to incorporate into their products. Alexa's popularity led to a worldwide boom in voice-enabled virtual assistants; smart speaker sales worldwide were predicted to reach more than $40 billion by 2024.

The Technologies

Creating a device that operates home appliances, surfs the Internet, integrates with other computing devices, and more required a number of technologies, including:

- **WIFI**: Echo connects to the Internet with its user's WIFI connection.
- **Microphones**: A seven-microphone array uses beamforming technology and noise cancellation to "hear" the user's voice even from across a room. Echo remains dormant until it hears the activating "hot word" or "wake word"—such as "Alexa"—that activates it to listen for commands or questions.
- **Bluetooth**: Echo can connect to other appliances and devices in the home using Bluetooth technology.
- **Cloud technology**: The device sends voice commands to a natural voice recognition service in the cloud called Alexa Voice Service, which interprets the commands and sends back the appropriate response.
- **Artificial intelligence**: The driving technology, AI, especially conversational AI, analyzes and responds to requests in near-real-time, and enables Alexa to converse with users.

Throughout its continued development of Echo and Alexa, Amazon has remained customer-centric, engaging with consumers to determine their wants and needs, and adjusting its designs accordingly.

For instance, when more than 40% of early Echo testers said they would mainly use the speaker for music, developers enlarged the device from hockey-puck-sized to one that holds a more powerful speaker.

However, Amazon founder, chairman, and CEO Jeff Bezos pressed his vision for the device to be more than a music player, encouraging Amazon's Lab126, which created Alexa, to develop the voice-activated assistant for other tasks, including shopping. Today, virtual assistants can open and display items for sale on users' screens and even make purchases. Echo and other virtual assistants also serve as a central hub from which users can operate smart-home appliances—another innovation that originated in Lab126.

The Benefits

Echo has enabled Amazon to not only remain relevant, but to position itself as a true innovator in the digital universe. Considered a novelty item first, Alexa may make

Amazon as synonymous with consumer computer, and Internet use as Google has been for searching and Microsoft Word for word processing.

Other benefits include:

- **Convenience for consumers**. Being able to get information, listen to music, watch videos, operate their homes, make calls, and perform many other tasks with a few words gives customers what they crave: instant gratification with the least effort.
- **Assistance for people with disabilities**. Voice-activated features empower people with disabilities to perform tasks they had previously found difficult or impossible, such as reading a book, or programming a thermostat.
- **Sales opportunities for Amazon**. Subscribers to Amazon's Prime premium service can use Echo to shop for groceries from Amazon's supermarket, Whole Foods Market, and have the groceries delivered to them. Amazon offers its Amazon Music Unlimited Service for Echo, by paid subscription. In shopping queries, Echo finds products sold on Amazon's website first.
- **Data collection**. Every command or request provides Amazon with information about Echo users, which enables the company to better tailor its services, products, and advertising to individual customers.
- **Partnership opportunities**. The company has teamed with hotels to place Echo speakers in its rooms, allowing customers to continue using Alexa while on the road. Amazon also allows other hardware makers to integrate Alexa into their products, further expanding the company's reach into more homes.

The Challenges and Lessons Learned

Latency, or lag time between command and response, was one of the first challenges in the development of Alexa. To store all the data needed to understand and process user requests takes one kind of memory; to analyze and respond quickly to those requests takes another kind. To make the experience as close to a genuine human interaction as possible, developers strived to minimize latency. They tried different kinds of processors until at last reaching a latency of 1.5 s—unheard of at the time.

Perhaps even more daunting a challenge was perfecting Alexa's responses so that she sounded human. Conversational AI is a complex and difficult technology; humans don't talk in a linear, logical fashion, but jump around, circle back, and refer to things said before—and context is key to getting it right.

To improve Alexa's comprehension and responses, its home screen displays text and graphics cards showing the user's recent interactions, providing an opportunity to provide feedback so it can learn. To work on speech recognition capabilities, Amazon hired people who had worked at a speech recognition company and bought two voice-response startup companies. It conducted numerous tests, over several years, to gauge which types of responses worked best. The ultimate goal is a natural, intelligent interaction that feels like speaking with a real person—one who knows you as well as you know yourself.

Case Study—Transportation, Travel and Tourism

Smart Tourism: Dubai

In 1960, Dubai was a rustic fishing village on the Persian Gulf. The discovery of oil in 1966 helped transform it over time into the most famous city in the United Arab emirates (UAE), with more than 2.2 million residents today. Its reserves were modest and so was oil production, however, tapering off in the early 1990s. Unfazed,

J. R. Reagan and M. Singh, *Management 4.0*, Blockchain Technologies, https://doi.org/10.1007/978-981-15-6751-3

the city's leaders shifted its focus to an alternate source of economic development: tourism.

The attract and accommodate visitors, the city built infrastructure that included three international airport terminals and many hotels, business parks, the world's largest shopping mall, and many other attractions. By 2015, Dubai had 15 million visitors a year from all over the globe, many of them deciding to stay: By 2018, 83% of its population were expatriates.

Like the proverbial shark that never stops moving forward, the city's government continued to forge ahead, establishing ambitious goals of leading the world in smart city technologies, including smart tourism initiatives designed to promote the city's new top-priority amenity: happiness.

Entering the Digital Age

Dubai tourism forecasts predicted continued increases to 20–25 million visitors annually by 2022, generating revenues of $7.6 billion a year. To meet these goals, the city appointed a Minister of Happiness whose job is to measure a variety of factors indicating the satisfaction of residents and tourists alike.

Smart Dubai, the city's public–private initiative using technology to provide "efficient, seamless, safe and impactful city experience for residents and visitors," has six pillars: Smart Economy, Smart Living, Smart Governance, Smart Environment, Smart People, and Smart Mobility. Projects comprise four components:

1. **Application**, using apps and dashboards to connect travelers to entities in the six pillars. Examples include Al-Fahidi, which provides virtual architectural tours of the city; Metro Moments, which informs visitors about points of interest on Metro stops; Dubai Tourism Apps, listing restaurants, landmarks, and other points of interest and providing a calendar of events; and in-house apps providing information to visitors of the city's major shopping centers.
2. **Service enablement**, involving analysis of data generated by these apps and other technologies for improvement of the traveler's experiences,
3. **Data orchestration**, ensuring the continual flow of data from devices for analysis, and keeping it secure, and
4. **Infrastructure**, installing, maintaining, and improving technologies for enhanced user experiences. The Road Transportation Authority of Dubai, for instance, connects users of six transportation apps on a unified platform to help move them from one place to another seamlessly and efficiently; the city's "Smart Palms," which resemble palm trees, provide free WIFI and device charging stations; and "Smart Tunnels" at its airports enables international to pass through immigration simply by walking through a tunnel, with no passport stamp or other human interaction.

Other technologies in place include:

- *Near-Field Communication*, which allows travelers to use their mobile phones to buy Metro and bus tickets,
- The Dubai Blockchain Strategy, promoting the use of blockchain technology to secure finance and banking data,
- A GPS navigator helping visitors to find their way around a city that has no street addresses.

To establish and maintain the plethora of technologies enabling its smart tourism initiative, the Dubai government has worked with a number of private partners including the TenCent International Business Group, the Chinese technology conglomerate behind WeChat and Weixin, which have more than 1 billion active monthly users, to enhance Chinese visitors' digital experiences in the city using artificial intelligence, the Internet of Things, cloud computing, and big data, and to collect and share tourist-generated data for Dubai's analysis.

So far, the city's focus on smart tourism appears to be yielding the desired benefits. Not only are visitor numbers growing steadily year by year, bringing it ever closer to its stated goal of being number one in tourism, but its "Happiness Meter" enables the

government to know whether it is fulfilling its top priority—its "Happiness Agenda."
Enabling users to click on one of three faces—frowning, neutral, and sad—the meter
interactively measures customer satisfaction with digital experiences.

"Through its simplicity and ease of use the Meter integrates seamlessly into the
experience flow of residents and visitors of Dubai," the government's website stated,
"while creating meaningful data towards the goal of making Dubai the happiest city
on earth."

Case Study—Social

Society 5.0—Japan

Japan is working on larger societal transformation plan where they will transform the Industry 4.0 to Society 5.0.

One question will be always present that is "**What better place to present a vision for a 'super-smart society' to the West, which you call Society 5.0 than Germany, the land of Industrie 4.0 or Industry 4.0?**" [1].

Note: Japan doesn't launch CeBIT's Society 5.0 vision, it already shared that vision in 2016.

As we know, Industry 4.0 is the digital transformation of manufacturing but in Society 5.0 will provide the solutions of range of problems beyond the digitizing the economy toward digitalization such as specific society problem in Japan.

J. R. Reagan and M. Singh, *Management 4.0*, Blockchain Technologies,
https://doi.org/10.1007/978-981-15-6751-3

That represents in CeBIT 2017 event with title "Society 5.0-Another Perspective," the role of technologies in super-smart society such as:

- The Internet of Things (IoT),
- Artificial intelligence (AI),
- Cyber-physical systems (CPS),
- Virtual reality (VR)/augmented reality (AR),
- Data analytics (big data),
- Blockchain technology,
- And many more.

A Digital Society for an Aging Population

One of Japan's unique problems is the aging population. While aging doesn't mean being a problem, there are challenges at various rates, and some of them can (and will) become less of a challenge as a result of smart solutions, allowed by technology but influenced by smart people.

Although the aging population is a problem for most nations, it is a problem for Japan in particular. As we discussed in our article on digital transformation in health care [2], Japan has 26.3% population age over the 65 years.

According to above-mentioned number, it is expected that more than 20% of the world population will be young. To put this number in perspective, it is expected till 2050 more than 20% of the world's population will be young for more than 60 years. Essentially, this means, all other countries can be follow the Japan Society 5.0 implementation methods and its impact in Japan 26.3% population. One of the Japanese society "Keindanren" [3] outline promote a diversified and flexible environment for each and every person where people do their job stress-free.

In health care, Society 5.0 will not only cure the chronic diseases of older people, but mainly the way health care is reorganized in light of this aging population reality, digital transformation efforts and the rethinking of care, including technology, are partly driven by the fact that people on average simply get older.

Society 5.0: Five Breaking Down Five Walls

The Society 5.0 has considered the health and other aspects of for Japan. It is not defined only for aging people problems such as mobility, live in practice, housing and so forth. Among the other problems that Japan is facing, natural disasters and pollution are same as other countries. To handle these obstacles, Society 5.0 community needs to deliver the way forward. For that, Japanese Society, Keidanren, has defined the Society 5.0 in five foundations.

However, just as in Industry 4.0, the fourth industrial revolution [4] Society 5.0 is also defined as an evolution in five social phases in the Keidanren position paper:

1. The hunting society,
2. The agrarian society,
3. The manufacturing society,
4. The information society,
5. The super-smart society.

What Japan is doing, in essence, is bringing the digitalization and transformation dimension—which is mostly bringing place at the level of individual organizations and parts of society—to the full national transformational strategy, policy, and even philosophy level. It's the furthest proposal we've ever seen in this regard. Below figure represents the evolutionary aspect of the Society 5.0 concept as introduced in the 5th Science and Technology basic plan of Japan.

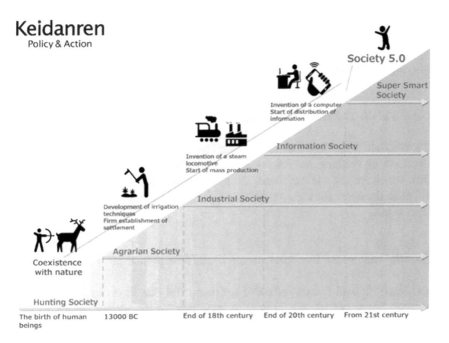

The Society 5.0 Evolution concept [5]

The 5 Walls to "Break Through" in Moving to Society 5.0

The 5 walls to break through in moving to Society 5.0 is describe at below.

The 1st Wall: Government and Policy Maker

The first wall is strategy designer of the nation that is government and their ministry. The governments formulate the national strategies and integration of government promotion system. It defined in Keidanren position paper.

The 2nd Wall: The Legal System

The 2nd wall is Legal system those develop and implement advanced technique according to the national law. That means government of country regulatory reforms and pushes the administrative digitalization.

The 3rd Wall: The Technologies

The 3rd wall of Society 5.0 is formation of the knowledge foundation. It completely depends on data management and all technologies to protect and leverage it, from cybersecurity to advanced technologies such as robotics, nano, and bio.

The 4th Wall: The Human Resources

The 4th wall of Society 5.0 will provide the educational reform, IT literacy, the expansion of available human resources with specializations in advanced digital skills which are just a few of them. The most interesting things in this wall is Japan will open the doors for highly qualified professionals in fields such as security and data science.

The 5th Wall: Social Implications and Social Acceptance

And final 5th wall is most effective and impactful due its acceptance boundary feature—Social Acceptance. It is most important for the society due its society-related aspect.

Obviously, in practice, Industry 4.0 and organizations overall will be major components in Society 5.0, yet it's not the industry alone: It's about all stakeholders, including citizens, governments, academia, and so forth.

If such a drastic shift in culture will succeed and the wall of social acceptance will be broken down is a question that will be answered in the future. To make assumptions in this regard would be Western ignorance on our part and a major mistake. So, who knows that? And is this a model that we can imagine in other parts of the world?

You can read more about all of this in the Keidanren outline, entitled "Towards the Realization of the New Economy and Society – Reform of the Economy and Society by deepening 'Society 5.0'" [8] and on this CeBIT [9], where the event was announced.

References

1. https://www.i-scoop.eu/industry-4-0/
2. https://www.i-scoop.eu/digital-transformation/healthcare/
3. Japan Business Federation @ http://www.keidanren.or.jp/en/profile/pro001.html
4. https://www.i-scoop.eu/Industry-4-0/fourth-Industry-Revolution/
5. http://www.keidanren.or.jp/en/policy/2016/029_outline.pdf
6. https://www.i-scoop.eu/information-management/document-capture-document-maging/
7. https://www.i-scoop.eu/information-management/
8. https://www.i-scoop.eu/cyber-security-cyber-risks-dx/
9. http://www.keidanren.or.jp/en/policy/2016/029outline.pdf
10. http://www.cebit.de/en/news/article/news-details-39442-503365.xhtml

Printed in the United States
by Baker & Taylor Publisher Services